THE
ULTIMATE
EXPERIMENT

THE ULTIMATE EXPERIMENT.

Man–Made Evolution

BY NICHOLAS WADE

WALKER AND COMPANY
NEW YORK

ISBN: 0-8027-0572-3

Library of Congress Catalog Card Number: 76-52575

Printed in the United States of America

Book design by Robert Barto

CONTENTS

Preface

MOLECULAR BIOLOGY has been called the art of the inevitable. The answers are all there: it is just a matter of looking them up in Nature.

But Nature yields its answers grudgingly, even to the massed legions of professional scientists who now pursue them. So it was something of an event, and perhaps not so inevitable, when there was discovered in 1973 a technique that serves as a kind of magnifying glass for reading DNA, the material in which the genetic instructions of living things are embodied.

The new technique enables the biologist to write in the genetic language as well as to read it. Indeed, the act of writing, it so happens, is an integral part of the reading. This fundamental duality of gene splicing, as I have called the technique, has an extended chain of consequences. Gene splicing is a means of intervening in Nature as well as interpreting it, of changing the world as well as understanding it.

The potency of gene splicing soon aroused apprehensions. The technique might be used carelessly, giving rise to novel organisms that would wreak havoc with human health or the environment, or it might in time lend itself to untoward practical uses, some perhaps malign, some well intended but socially disruptive.

In the chapters that follow I describe the development and discussion of these concerns and the means devised to address them. Some conclusions are offered, as well as enough fact, I hope, for the reader to draw his own.

The bulk of this book was written in February and March 1977, as the first round of the debate over gene splicing was coming to a close. Except where otherwise mentioned, it draws on my reporting of the issue for the news section of *Science*, a weekly scientific journal published by the American Association for the Advancement of Science.

I thank Richard K. Winslow of Walker and Company for suggesting that the book be written. Roy Curtiss, professor of microbiology at the University of Alabama; Benjamin Lewin, editor of *Cell*; and Colin Norman, Washington correspondent of *Nature*, were kind enough to read the draft of the book at short notice. I thank them and other friends for useful criticisms.

—Nicholas Wade

THE
ULTIMATE
EXPERIMENT

1

THE KEY
TO THE KINGDOM

A TURNING POINT has been reached in the study of life. A turning point of such consequence that it may make its mark not just in the history of science but perhaps even in the evolution of life itself.

The turning point is the discovery in 1973 of a technique for manipulating the stuff of life. Known at present by the awkward name of recombinant DNA research, the technique is in essence a method of chemically cutting and splicing DNA, the molecular material which the genes of living organisms are made of. It enables biologists to transfer genes from one species to another, and in doing so to create new forms of life.

All previous speculations about genetic engineering can now be forgotten. The new gene-splicing technique is far more potent and dexterous than anything the theorists contemplated. Its impact will be profound, and it will be felt almost immediately.

For the last twenty-five years molecular biology, the study of DNA, has made the fastest progress of any branch of science. A mass of pure knowledge has been accumulated, but so far with rather little practical effect. Gene splicing promises to offer a quick bridge between this vast untapped storehouse of pure knowledge and the practical world of medicine, agriculture and industry. Everything that biologists have learned in their attempt to understand nature now becomes a means to change it.

Some thirty-five years ago physicists learned how to manipulate the forces in the nucleus of the atom, and the world has been struggling to cope with the results of that discovery ever since. The ability to penetrate the nucleus of the living cell, to rearrange and transplant the nucleic acids that constitute the genetic material of all forms of life, seems a more beneficent power but one that is likely to prove at least as profound in its consequences.

It could well prove comparable to that other biological revolution in man's history, the domestication of plants and animals. That achievement, by the people of the Neolithic Age, opened a doorway for man to pass from uncertain existence as a hunter and gatherer to life as a farmer, herder, and city dweller. From that beginning some seven thousand years of urban civilization have followed. Yet Neolithic man, like animal and plant breeders ever since, did not create new species; he only selected, and reinforced by breeding, the characteristics he desired from among those already within the natural genetic potential of a species.

Scientists today cannot design entirely new genes any more than Neolithic man could (although that may eventually be possible). What the new gene-splicing technique does make possible is the transfer of genes from one species to another, regardless of the reproductive barriers that nature has built between them to isolate one species from another. It is now becoming technically possible (though practically fruitless) to intermingle the genes of man and fungus, ant and elephant, oak and cabbage. The whole gene pool of the planet, the product of three billion years of evolution, is at our disposal. The key to the living kingdom has been put into our hands.

There are occasional suggestions, made on scientific or moral grounds, that the key should be thrown away. Such abnegation of intellectual curiosity is not in man's nature, and in any case the question is moot: the door to the treasure-house is already ajar, and the only question remaining is what use will be made of the riches within.

The immediate benefit of gene splicing and the kindred techniques it generates will be pure knowledge. The technique gives biologists a handle on many fundamental problems, some of which had seemed almost insoluble by previous methods. How genes work in human and other cells, how they are switched on and off, how they obey the master plan that guides the development of each organism from egg to adult—these are some of the problems in which the pace of discovery will now appreciably quicken.

It is reasonable to suppose that the accumulating body of knowledge about the cell will eventually provide an understanding of diseases such as cancer, which in turn may perhaps (but not necessarily) lead to appropriate methods of treatment.

Such possibilities lie a long way off. Much closer at hand is the use of gene splicing to program bacteria to produce valuable proteins of use not just in research but also in medicine and industry.

The basis of the proposed method is to grow and harvest bacteria much in the manner used to produce substances such as penicillin. But instead of the bacteria producing a substance specified by their own hereditary information, a new gene is spliced into them so that they synthesize whatever substance the gene specifies.

Though not yet proved, the concept should offer a powerful means of producing many kinds of important protein, such as insulin, interferon (the body's antiviral substance), and vaccines.

Further down the road is the possibility of using the

technique to modify whole organisms. Plants, since they can be grown from a single cell, are a prime candidate for genetic engineering. Most important crop plants cannot extract nitrogen from the air; they must have it supplied to them in the costly and polluting form of nitrogen fertilizer. Were it possible to transfer to wheat the relevant genes of the nitrogen-fixing bacterium, a new agricultural revolution would be in the making.

If genes can be added to plants, why not also to humans? Many people suffer from genetic diseases, such as sickle-cell anemia. Gene therapy is the word already given to the idea that it may one day be possible to switch off the defective gene and switch on or insert the genes that specify the correct product.

Such projects are only gleams in the experimenter's eye, but they are indicative of the hopes being held out for the new technique. Applications that no one has yet thought of will doubtless turn out to be at least as significant as those now being discussed. The gene-splicing technique is only the first of a succession of technologies that will make possible ever finer control of the chemical materials of life. Other technologies developed in the course of civilization are merely extensions of man's hands or senses. The ability to manipulate the stuff of life is an art of a different order, the ultimate technology.

All technologies have their unintended and untoward side effects, and a technique of such power as gene splicing is unlikely to prove an exception.

The most immediate fears concern its possible threat to laboratory workers and, through them, to the public health. In transferring genes from one organism to another, might not a researcher unknowingly enhance the virulence of an existing microbe or perhaps create a new pathogen for which no defenses had been prepared?

Though far from likely, such a possibility is made more tangible by the fact that the standard organism of laboratory study is *Escherichia coli,* a bacterium whose natural habitat is the gut of humans and other warm-blooded animals.

Moreover, many of the scientists who will be manipulating genes in laboratories throughout the world are molecular biologists, not all of whom are trained in the painstaking procedures used by microbiologists to prevent infection of themselves or others.

The gene-splicing technique thus raised from its inception a question of laboratory safety and public health. The past three years of intense debate, first among scientists and then in public forums, have focused almost exclusively on this specific problem and the means to cope with it.

More recently a second kind of doubt has been raised about gene splicing: the possibility that the technique may have long-term evolutionary consequences. Single-celled organisms like bacteria are organized on somewhat different principles from those that govern the cells of higher plants and animals. Transferring animal or plant genes into bacteria, the argument goes, might endow bacteria with the genetic signals of higher cells and render the animal and plant kingdoms vulnerable to a new mode of bacterial attack. Whatever the merits of the argument, and they are hotly disputed, it would seem in a purely general way that in shuffling genes from one organism to another, scientists are playing evolution's game without exactly knowing either the rules or what the forfeit may be for transgressing them. But many biologists believe, and they may well be right, that evolution's rule is that anything goes and that thus there is nothing to be worried about.

Accidents apart, there will always be the chance of putting so potent a technology to malign use. It was for lack of just such a technique as gene splicing that the U.S. Army's twenty-five-year program in biological warfare produced so little of military value. The United States voluntarily renounced offensive biological warfare in 1969, and both the United States and the Soviet Union are signatories of a convention that prohibits the development, production, and stockpiling of biological weapons. Though it does not preclude research, the convention is a significant barrier to the military exploitation of gene splicing, at least by these two superpowers. Whether or not it will inhibit smaller nations seeking to develop a cheap substitute for nuclear terror is another matter. As gene splicers' capabilities increase, there is no knowing what temptations may be presented to military planners or those who perceive themselves in desperate straits.

Gene splicing is so simple a technique that for most present purposes it requires only a few dollars worth of special materials, all commercially available, and access to a standard biological laboratory. Political analysts worry about the proliferation of nuclear weapons and the vulnerability of peaceful nuclear materials to terrorist diversion. Gene splicing seems a far more beneficent technology; yet should it prove otherwise, there will be no way of preventing general access to it.

Terrorists will doubtless continue to prefer the multitude of conventional weapons at their disposal. Perhaps more to be feared is the deranged do-gooder who decides to take some unilateral action for what he conceives to be the relief of suffering humanity. Figuring that many deaths from lung cancer could be prevented by the eradication of the tobacco plant, a scientist or competent technician might use the technique to improve upon the natural virulence of the tobacco plant's viruses.

An increasingly potent art is about to be placed at the disposal of every government and several thousand biologists. Even if perverse uses are avoided, the inventions that flow from it will certainly require some hard decisions. Gene splicing is a significant first step toward making human genetic engineering technically feasible. Controlling our progress down that route will not be simple. The alluring benefits of genetic engineering, it can be envisaged, may be established first in crops, then in domestic animals, next in remedying human genetic diseases, and then in enhancing natural growth so as to ensure that each individual attains his full genetic potential. Somewhere along that route, the engineers may cross, perhaps imperceptibly, the Rubicon that everyone had supposed to be the natural stopping point, the inviolability of the human genetic constitution. The way would then be opened for *Homo sapiens* to bring to birth his finest creation: *Homo sapientissimus.*

Such a chain of events would probably be broken at an early stage by technical infeasibility or social resistance. In any case, society is far from incapable of both reaping the benefits of a technology and minimizing its risks. Yet some technologies seem to possess an inherent pace of development, which may not necessarily be as deliberate as society would prefer.

Gene splicing, at any rate, is a technique that will profoundly transform our knowledge of ourselves and our evolutionary history. In doing so, it will bring about a new biological technology, whose range of application can at present only be guessed at but which could eventually extend to changing humankind as well as nature. In what follows is described the reactions among scientists and the public to the birth and first stirrings of the ultimate experiment.

2

LIFE AS THE EXPRESSION OF DNA

OF ALL THE branches of human inquiry, none is advancing faster now, or moved more haltingly in the past, than the study of biology and its central realities, the mechanisms of evolution and heredity.

The belief that species of plants and animals had remained fixed and unchanging since time immemorial lingered on for many centuries after biblical ideas about astronomy had been challenged and abandoned. It was scarcely more than a century ago, in 1859, that Darwin proposed in his theory of evolution that the species are not constant but are gradually modified by the forces of natural selection; in his view, the fittest individuals in each generation survive to become the parents of the next; and thus over time, through the pressures of adjusting to natural changes in its environment, each species gradually evolves into a slightly different species, or else passes into extinction. Acting over millions of years, even the gradual forces of natural selection, Darwin supposed, are potent enough to engender the great variety of the planet's different life forms, creatures as dissimilar as elm and hawk, man and moth, whale and worm.

Darwin did not understand the source of the variation upon which natural selection worked. That was to become apparent much later, after the development of the pivotal concept of genes. Darwin's theory was bitterly criticized in his lifetime; a worse fate, neglect, befell the work of the abbé Gregor Mendel, which was published in 1865 and all but ignored until the beginning of this century. Mendel realized from his breeding experiments with peas that the range of variation in a plant's progeny was not always continuous and that certain traits, such as seed color and appearance, were either inherited or not inherited in a particular pattern. He inferred that the traits were determined by discrete hereditary factors. These discrete factors are now called genes.

Combining studies of heredity with what could be seen under the microscope, biologists came to understand that genes must be arrayed in linear fashion on the threadlike structures known as chromosomes, which exist in every living cell. When a cell divides, each of its chromosomes duplicates and splits, so that each daughter cell receives a full set of the chromosomes and the genes they carry.

The genes of each organism constitute its hereditary blueprint, and each cell possesses a complete copy of the blueprint. What makes a liver cell different from an eye cell or brain cell is presumably that in each sort of a cell a particular group of genes is active, while the rest are somehow switched off.

It took biologists nearly another century after Mendel to discover what genes are and how they determine the nature of the cells and whole organisms, which are their physical expression.

The first step toward answering this fundamental question was taken only in the 1940s, when it was shown that genes are associated with a particular kind of molecule known as DNA, or deoxyribonucleic acid. Most biologists at the time had reasons for believing that DNA could not be of any significance in heredity. Nevertheless, a few chemists went to work on it and established that, despite the molecule's great size and complexity, it is composed of a large number of very similar chemical building blocks, of which there are only four different sorts.

These building blocks are known to chemists by the general name of bases (they are basic rather than acidic), and the four different sorts of base are designated A, G, T, and C (after their full chemical names of adenine, guanine, thymine, and cytosine).

The cells of different organisms may possess widely different quantities of DNA, but one striking constancy was discovered by DNA chemist Erwin Chargaff of Columbia University. Whatever the source of DNA molecules, whether they came from man, mouse, or microbe, all contained the same number of A's as T's and the same number of G's as C's.

That was one vital clue to solving the structure of DNA. The other was an X-ray photograph of DNA crystals, taken by the late Rosalind Franklin of the University of London, which indicated that the molecule had some kind of spiral structure.

The problem suggested by these and other clues was solved not by professional biologists, but by two young scientists working in the physics department of the University of Cambridge in England. James Watson and Francis Crick made the landmark discovery that linked the observable facts of heredity with the atoms and molecules in which they are grounded.

Their discovery was that the DNA molecule consists of a pair of chains, one of which is a special kind of replica of the other.

The chains lie a fixed distance apart from each other, with their bases interacting. The size and shape of the four kinds of bases is such that where one chain has base A, the opposite base on the other chain has to be T, and vice versa. Also, where one chain has G, the only base that will fit opposite it is C, and vice versa.

This relationship means that the sequence of bases along one chain of a pair determines what the sequence must be on the other. Where one chain reads A-G-G-C-C-C-T-T-T-T, the other chain has T-C-C-G-G-G-A-A-A-A:

$$-A-G-G-C-C-C-T-T-T-T-$$
$$-T-C-C-G-G-G-A-A-A-A-$$

The two chains, which are not straight but coiled in the shape of a spring, or double helix, embody a perfectly appropriate system for conserving and for copying the hereditary information. When the double strand is pulled apart and the new chains built up alongside the old, the order of the incoming bases is determined by the bases in the existing chains—A's matching to T's, G's to C's, and vice versa. Thus each new chain is a copy (although a negative copy) of the old chain, and the two new double chains are perfect copies of the parent DNA.

Watson and Crick published their discovery of the structure of DNA in 1953, and much of what biologists have done since has been an attempt to follow up the consequences of that finding. For it seems to be that all forms of life, all the observable facts of heredity and evolution, are only the physical expression of information encoded in DNA molecules.

That information is encoded by the sequence of the bases along the DNA chain. The bases constitute a four-letter alphabet known as the genetic code, which, as far as is known, is universal to all forms of life on earth.

A gene is an instruction written in this alphabet. A gene is a length of DNA that includes on the average about one thousand pairs of bases.

The grand principle of life—the central dogma, as

biologists call it—is that DNA makes RNA, RNA makes proteins, and proteins make everything else.

RNA, or ribonucleic acid, a molecule very much like DNA, is the material used by the cell to make copies of gene-length segments of DNA. From the RNA copies of genes on the DNA, the cell receives instructions on what proteins to make and how to make them.

Proteins, like RNA and DNA, are linear molecules. Their subunits are a family of chemicals known as amino acids, of which there are twenty different kinds.

A gene is said to be "expressed" when the protein it specifies is manufactured by the cell. Expression starts when the stretch of DNA that constitutes the gene is copied onto a stretch of RNA. Then the RNA molecule passes like a tape through the cell's protein-synthesizing machinery. As it does so, the order of its bases determines the nature and sequence of the amino acid subunits that are to compose the protein, and the cell's machinery zips together the lined up subunits into a complete protein molecule. The average size protein consists of about three hundred amino-acid units strung together in a chain, the chain being folded into a specific shape that is characteristic of each kind of protein molecule.

Proteins vary widely according to the nature and sequence of their amino-acid subunits. Because of this versatility, the proteins developed over the course of evolution are able to perform an enormous range of different biological functions. There are structural proteins, which form muscle and cartilage. Proteins such as hemoglobin carry the gases in the blood, while others such as insulin, serve as hormones. The largest class of proteins are biological catalysts, known as enzymes, which control and direct the detailed biochemistry of the body. For each of the thousands of chemical reactions necessary to the body's function, a specific protein-enzyme exists to direct and control it.

How did different proteins develop? The copying of

a DNA molecule, which occurs when a cell divides, is an exact but not invariably perfect process. Just occasionally a wrong base might be incorporated in the DNA. The new base might change the amino-acid subunit coded for at that point in the gene. Substitution of one amino acid for another in a single protein may not seem like a major change, but that is all, for example, that distinguishes the hemoglobin of sickle-cell anemics from ordinary hemoglobin.

Most such changes, known as mutations, are harmful, and the organisms possessing them do not survive. But an occasional beneficial mutation confers an advantage on the individuals in which it occurs and so has a chance of being retained. (Sickle cell anemia, for example, has a compensating advantage—resistance to malaria.)

Substitution of one base for another is the smallest kind of mutation, but grosser mutations may occur, such as when a whole length of DNA is copied twice over by mistake. Such an error would endow an organism with two copies of the same gene. In subsequent generations one copy might continue to serve its proper function while the other would be free to accumulate mutations until, by rare chance, it produced some new protein of use to the organism. Thus a new gene would later enter the gene pool of the species.

Very occasionally, the whole gene set of an organism might be duplicated. An animal with an extra set of genes would have vast spare potential upon which natural selection might exert improvements. In terms of quantity of DNA, there is not so great a gap between a bacterium and a human. About ten doublings of the genetic material will do it. The first single-celled organisms to appear in the earth's fossil record lived about 3.5 billion years ago. Assuming they possessed one fourth as much DNA as do present-day bacteria, it would take only one doubling error every 300 million years to account for the increase in the content of DNA between these first single-celled organisms and man.

Errors in the copying of DNA, from single-base mutations to doublings of the whole gene set, are the basic source of the variability upon which natural selection works and subtly shapes new species from old. The fusion of Darwin's theory with Watson and Crick's discovery has provided a remarkable insight into the evolutionary history of life. The insight may never be powerful enough to penetrate the mists that surround the dawn of life on earth, but it has already made comprehensible at least an outline of what may have happened.

The earth was shaped from accreted debris in orbit about the sun; its consolidation as a planet is dated (by study of residual radioactivity in rocks) to about 4.5 billion years ago. From the water, minerals, and gases in the primitive atmosphere a variety of complex chemicals were formed through natural processes, one of which was a DNA molecule—or something like it—just a few bases long.

In the primitive chemical soup the first DNA molecules would often have been buffeted apart into separate chains, on which new chains of bases could be formed to create new molecules. But perhaps in this cycle of purely chemical replication there was room for error, and a wrong base—perhaps one that was in greater abundance in the environment—became incorporated into the DNA. In a living organism such a change would be called a mutation. The word is not inappropriate because even the simplest DNA molecule, though just one of a myriad of chemicals on the early earth, would have possessed the two essential characteristics of life—the capacities of replication and mutation.

Once a chemical system had emerged in which the

hereditary transmission of mutable information could occur, the processes of natural selection could go to work, pushing the system toward constant improvement, ever greater capacity and complexity. The primitive DNA molecules somehow developed the association with proteins that is expressed in the genetic code. Or perhaps it was the proteins, evolving first, which developed the association with DNA. In either case, the early DNA-protein systems then derived an advantage from working within a closed envelope; that would have been the genesis of the primitive cell.

These first steps were doubtless the least probable of all the historical events that attended the emergence of life on the earth, which might be why the process took a billion years. But after the appearance of the first primitive cells some 3.5 billion years ago evolution quickened its pace, progressing from single cells to multicellular organisms; throwing up a vast diversity of creatures that swam, flew, and walked; experimenting with one design after another and discarding each almost as soon as it was created.

To personify evolution is a shorthand way of referring to the process of natural selection and does not imply conscious design or intention. There is no necessary reason to suppose that anything other than the laws of physics and chemistry governed the emergence of life and its subsequent evolution.

Yet living cells, and the DNA molecules of which they are the expression, are much more than just a bunch of chemicals whose reactions are governed by the laws written in scientific textbooks. That extra factor is not some mysterious vitalist spirit, but neither is it something which is completely understood: it is history.

Every DNA molecule that now exists in nature is the result of some 3.5 billion years of evolutionary history, the product not just of the laws of physics and chemistry but also of time and chance. Unlike formaldehyde or benzene or DDT, every DNA molecule has a past. In this sense at least, DNA is not just another chemical. Discovery of the double helix in 1953 revolutionized biology. It unified and made explicable a vast range of diverse observations, from the biochemistry of the cell to the workings of evolution. But filling in the details of this grand explanatory framework has proved a slow and arduous task. Some considerable progress has been made in understanding bacteria. But the greatly more complex cells of animals and plants, if not exactly virgin territory, have been a jungle too thick and interwoven to be penetrated with the available analytic tools. That was the case, at least, until the invention of the gene-splicing technique.

3
GENE SPLICING
FOR BEGINNERS

THE GENE-SPLICING technique is so important to researchers because of the intractability of the problems they had begun to come up against. Biologists have only recently started to acquire a detailed understanding of viruses, entities so simple that they can scarcely be considered living organisms. They are even further from understanding the primitive order of single-celled organisms known as procaryotes, which include bacteria and certain algae. Even less well understood than procaryotes are eucaryotes, the cells which comprise the bodies of multi-cellular organisms such as higher plants and animals.

The most intensively studied of any procaryotic organism is the laboratory strain of *Escherichia coli*, the human gut bacterium. *E. coli* possesses a single chromosome, which consists of about four million base pairs, enough to encode some four thousand genes. According to an estimate by James Watson, we know perhaps a third of the metabolic reactions which take place in *E. Coli*, and it may take another twenty to twenty-five years before we approach the state of knowing them all.[1]

Eucaryotes, the other great order of cells, which includes those of animals and higher plants, are vastly more complex than those of procaryotes such as *E. coli.* Human cells are about six hundred times the size of *E. coli.* They possess not one chromosome but forty-six. Each human cell contains about a thousand times more DNA than does *E. coli.* The methods of analysis which have proved so successful with bacteria are much less powerful when applied to eucaryotic cells. The general approach of some of these methods is to shatter the mechanism of the cell and hope to deduce from each piece of debris how it worked in life.

Gene splicing replaces the hammer with a scalpel and magnifying glass. The researcher can now chop up the DNA into precise pieces of manageable size and then grow up multiple copies of each piece until there is enough for chemical analysis.

Two key biological tools are needed for gene splicing: restriction enzymes and a cloning vehicle. Restriction enzymes are the scalpel part of the technique. Produced by bacteria, the property of the enzymes is to cut DNA molecules. But they do not do so at random. They "read" the bases of DNA molecules and snip the double helix only when they chemically recognize a particular sequence of bases.

Different restriction enzymes cut at different sequences. An enzyme produced by *E. coli*, for example, known as Eco RI, cuts DNA wherever the base sequence on one strand reads -G-A-A-T-T-C- and that on the other is the complementary sequence -C-T-T-A-A-G-.

An important feature of the sequences recognized by certain enzymes is that they occur in DNA molecules neither very often nor very seldom. The Eco RI cutting point, for example, occurs once every four to sixteen thousand base pairs. The enzyme thus cuts the DNA molecules into fragments which may contain roughly from four to sixteen genes each.

Here at once is a tool for cutting the immensely long DNA molecules of living organisms into a pattern of smaller pieces. A particular enzyme can be relied upon to produce the same pattern of fragments. Use a different restriction enzyme, and a new pattern of pieces will result.

(Why do bacteria produce restriction enzymes? Apparently as a defense against the DNA of invading viruses. Why don't the restriction enzymes attack the bacterium's own DNA? The bacterium always produces a twin for each restriction enzyme, known as a modification enzyme. The modification enzyme recognizes the same DNA sequences and chemically modifies those on the bacterium's DNA—presumably it gets there first—so as to camouflage the sequence from the restriction enzymes. It is fortunate for researchers that bacteria produce restriction enzymes; the design of such proteins is quite beyond biologists' present powers.)

The second major step in gene splicing is to clone the piece or pieces of DNA produced by the restriction enzymes. As the restriction enzyme is a scalpel, so the cloning technique becomes the magnifying glass. A clone is the name given to a population of identical bacteria, all of them descended by cell division from a single individual. *E. coli* takes only twenty minutes to divide into two identical daughter cells, which in turn divide into four cells, then eight cells, sixteen cells, thirty-two cells, and so forth in geometric progression until after a short period many millions of copies have been produced of the cell that founded the clone. Gene splicing takes advantage of this multiplicative process by inserting into the founding bacterium the DNA fragment under study.

But the fragment must be inserted in such a way that it multiplies each time the bacterium's own DNA multiplies. One way to accomplish this is to join the fragment to the DNA of a virus that invades bacteria. Most viruses consist of a short piece of DNA wrapped up in a protein coat. Some viruses, once they have penetrated their target cell, have a trick of shedding their coat and inserting their own DNA into the DNA molecule of their host. There the virus lies latent, being transmitted to all the cell's progeny just as if it were one of the cell's own genes, until some triggering mechanism induces the DNA to become a virus again.

If one of the DNA fragments generated by a restriction enzyme is added to the DNA of this kind of virus, the fragment will get integrated into the bacterium's chromosome and be replicated along with the virus during the formation of a clone. The virus in this case would be serving the role of a cloning vehicle for the inserted fragment of DNA.

A generally more convenient cloning vehicle, however, is an entity known as a plasmid. Plasmids are small pieces of DNA that are carried by certain bacteria as extra minichromosomes. They carry a dozen or so genes, exist separately from the bacterium's main chromosome, and replicate independently at cell-division time, with a copy going to each daughter cell. Some plasmids can also be transferred from one bacterium to another during mating.

A plasmid of particular importance in gene splicing is one known as pSC101 (p for plasmid, SC for Stanley Cohen, in whose laboratory it was obtained).* The plasmid is small and contains only a few vital functions. These are (1) the replication site, the sequence of bases that signal to the bacterial enzymes that the plasmid is to

---

*The origin of pSC101 is at present veiled in doubt. When first isolated, it was believed to have been derived from a larger, ring-shaped plasmid known as R65 which had supposedly lost a segment of its DNA and re-circularized to form the smaller, pSC101 ring. It now appears that pSC101 shares only 25 percent of its DNA sequence in common with its presumed parent. This suggests either that it was derived from R65 by some complicated process which masks their direct relationship, or that it is in whole or part a contaminant. Cohen is re-examining the ques-

be replicated, (2) a gene which confers resistance to the antibiotic tetracycline, and (3) one and only one base sequence that is recognized by the restriction enzyme Eco RI.

Plasmid DNA happens to exist not as a linear molecule but in the form of a ring. Thus, when pSC101 is attacked by Eco RI, it is not cleaved into two pieces, it is simply converted from a ring to a straight molecule.

The next step is to link the opened-up pSC101 to the DNA fragment being cloned. The linking is made possible by another remarkable feature of some restriction enzymes; when they cleave the double helix, they do not make a clean cut but snip one strand a few bases further down than the other, leaving what chemists call "sticky ends." Eco RI, for example, cuts between the G and A in its target base sequence, but since the G-A groups on each strand are not directly opposite each other, the effect is to make a staggered cut:

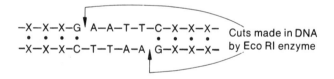

Cuts made in DNA by Eco RI enzyme

The reason why chemists call the protruding A-A-T-T sequence a sticky end is that DNA bases have a chemical affinity for their opposites (A for T, G for C). A DNA fragment created by cutting a molecule with Eco RI will spontaneously anneal, or recombine, to any other frag-

---

tion but has not reached any definite conclusion, except to say that pSC101 could not be derived from R65 by simple recircularization, as at first supposed. Even if pSC101 turns out to be related to some other plasmid which may have contaminated the R65 culture, Cohen considers that no biohazard issue would be raised because of the free circulation, and equally unknown origin of, all enteric plasmids. From the purist point of view it would be better to know than not to know the identity of the most widely used cloning vehicle in gene splicing research.

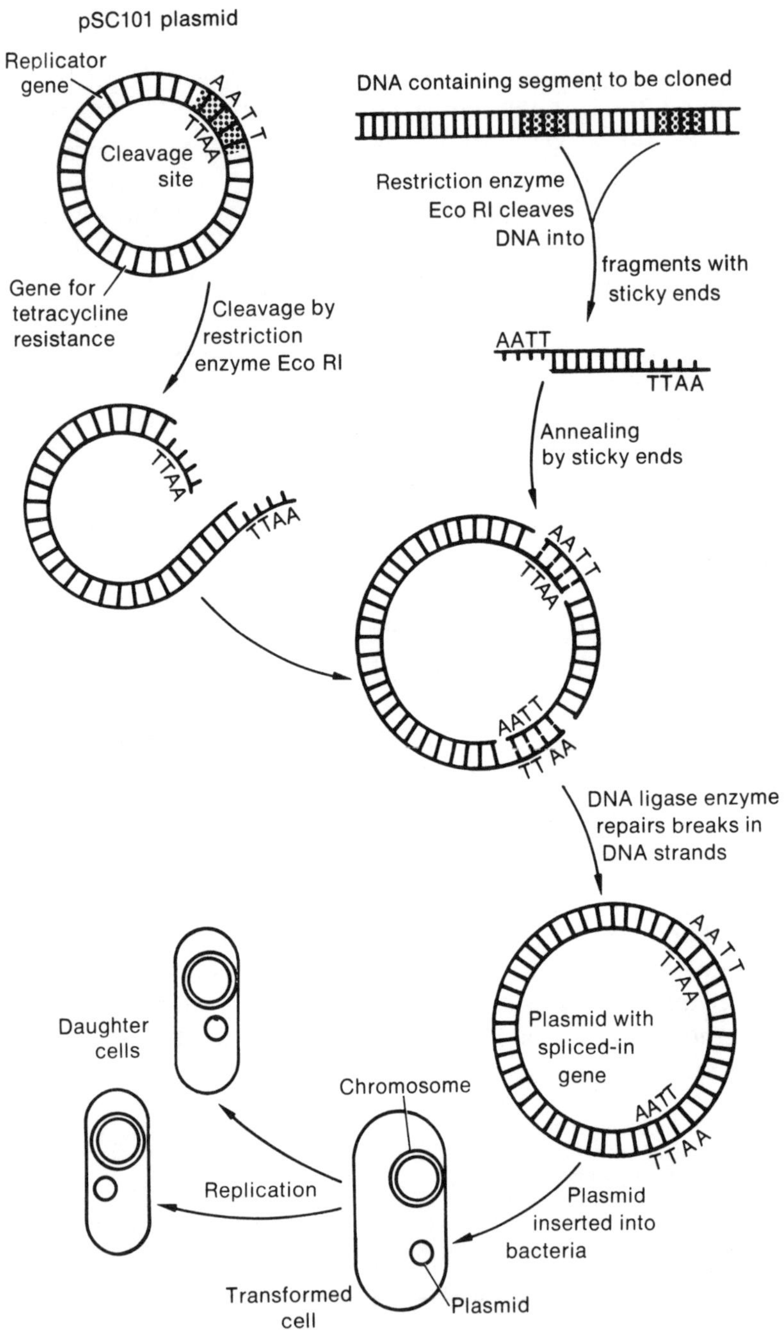

FIGURE 1. THE GENE-SPLICING PROCEDURE. (Adapted from Stanley N. Cohen, "The Manipulation of Genes," *Scientific American,* July 1975, p. 30.)

ment that has been created by the same enzyme and has the same kind of sticky end—hence the phrase *recombinant DNA*. Take any Eco RI–generated DNA fragment, mix it with pSC101 molecules broken open with Eco RI, and the sticky ends of the fragment will anneal with the identical sticky ends of the plasmid.

This, then, is the way in which the pSC101 cloning vehicle is charged with the DNA fragment to be cloned (see fig. 1). But even when the A-A-T-T ends of the fragment and plasmid have annealed with each other through the attraction between their sticky ends, there are still two breaks in the main chain of the plasmid ring. These can be repaired by mixing into the test tube another kind of enzyme, called DNA ligase, whose natural function is to mend breaks in DNA.

The genes or DNA fragment to be cloned are now an integral part of the plasmid. The next step is to insert the plasmids into bacteria. A broth of *E. coli* K12, the standard strain, is dosed with a salt and the plasmids are mixed in. The flask of broth is then transferred from a bed of ice to warm water. Under the shock of the salt and the sudden temperature change, the cell walls of the bacteria become porous, allowing the plasmids to slip inside.

Not all of the bacteria in the broth will be invaded by a plasmid, and those that are not are neatly eliminated by the addition of the antibiotic tetracycline to the flask. Since pSC101 carries the gene for resistance to this particular antibiotic, the bacteria that contain plasmids are unaffected by it and those that don't are killed.

The bacteria are now poured out onto a plate of nutrient jelly, on which each cell starts dividing to form a clone of identical progeny cells. Every cell in the clone will contain a pSC101 plasmid with its inserted DNA fragment, so that the DNA has now multiplied by as many cells as there are in the clone.

The researcher can now study the DNA fragment as it is, to learn what genes it contains and what proteins they code for, or he can harvest the fragment and its protein products in pure form for further tests.

Before gene splicing, the eucaryotic cell was a million-gene system almost too complex for serious analysis. With the technique almost any gene (or at least the DNA fragment containing it) can be snipped out, cloned, and manufactured in pure form.

The technique was invented by Stanley N. Cohen and Annie C. Y. Chang, who work at the Department of Medicine in the Stanford University School of Medicine, and Herbert W. Boyer and Robert B. Helling of the Department of Microbiology at the University of California at San Francisco (Helling is now at the University of Michigan). The technique draws upon the work of other scientists, such as those who discovered how restriction enzymes work, but it was these four who put it all together. They published the details of their invention in 1973,[2] and so initiated the age of synthetic biology exactly twenty years after the great analytic phase that followed the discovery of the double helix.

The power of the new technique was fascinating but also a little troubling to its inventors. Until then nobody knew if genes from one organism would function or even survive in another. Cohen and his colleagues began to discover that with their technique these biological barriers could be crossed very easily. First, Cohen and Chang took a gene from an unrelated bacterium known as *Staphylococcus aureus* and inserted it on the plasmid into *E. coli*. The gene was one that conferred resistance to penicillin. They proved it was working in its new host because the *E. coli* became resistant to penicillin.

They and other colleagues next performed a landmark experiment in biology by trying to jump over an even larger biological barrier. On Friday, July 27, 1973, they spliced into plasmids genes from *Xenopus laevis*, the South African clawed toad. During the following week the toad gene-carrying plasmids were inserted into *E. coli*. Despite the enormous evolutionary distance between the two organisms the toad's genes were reproduced by the bacterium.

But the success of the technique raised some hazardous possibilities. When Cohen and Chang inserted the penicillin resistance gene into their laboratory strains of *E. coli*, they were careful to choose a type of resistance that natural populations of *E. coli* have already acquired. But other researchers, should they have been less thoughtful, might have introduced antibiotic-resistant genes into bacteria that did not naturally possess them.

Many biologists are studying viruses that cause tumors in animals; with the new technique it would be a great temptation to insert tumor virus genes into *E. coli* to see how the genes exerted their cancer-causing effects. But what if such extra genes should somehow make *E. coli* itself a harmful, even a cancer-causing, organism?

Another procedure of great interest, which the toad gene transfer demonstrated would be technically feasible, is what has come to be known as the shotgun experiment. This involves taking the entire DNA of an organism—of a fruitfly, say—chopping it up into fragments with a restriction enzyme, and loading all the fragments onto plasmids for cloning in the usual way. For the researcher, the conceptually elegant result is that the fruitfly's total gene set, some sixteen million bases in total length and otherwise too complex for serious analysis, now lies in manageable pieces, with each piece amplified to useful quantities inside a clone of bacteria. Yet a possible hazard of the shotgun experiment is that one of the many thousand genes being shotgunned may in fact be harmful and could endow *E. coli* with its harmful propensities.

No one had given much thought to such possibilities

in 1973, and it was hard either to confirm or dismiss them. But it was at least evident that the power of the technique might be misused by researchers who failed to consider the consequences. Besieged with requests for pSC101 from other scientists, Cohen and Chang therefore asked for assurances from recipients of the plasmid that they would not use it to introduce either tumor viruses into bacteria or genes for antibiotic resistances that the bacterial species did not already possess in nature. Also, in order to keep track of the distribution of pSC101, they asked their colleagues not to pass it on to other laboratories.[3]

Thus, from its very inception, gene splicing was marked by an awareness of its hazards on the part of the researchers and by an attempt to take appropriate precautions. But the warning issued by Cohen and Chang, whatever its adequacy to the risk, did not allay the concerns raised by the protean power of the device they and their colleagues had invented.

NOTES

1. James D. Watson, *Molecular Biology of the Gene*, 3rd ed. (W. A. Benjamin, Inc.: Menlo Park, Calif. 1976): p. 81.

2. Stanley N. Cohen, Annie C. Y. Chang, Herbert W. Boyer, and Robert B. Helling, "Construction of Biologically Functional Bacterial Plasmids *in Vitro*," *Proceedings of the National Academy of Sciences* **70** (November 1973): 3240–3244.

3. Stanley N. Cohen, "The Manipulation of Genes," *Scientific American*, July 1975, p. 32.

4
ORIGINS OF THE MORATORIUM

THE FURIOUS DEBATE over safety that was to be triggered off by the invention of gene splicing was in some respects long overdue. Other powerful biological techniques, such as fusing the cells of different species or producing mutations in viruses and bacteria, could perhaps have raised equally serious doubts but they passed into common laboratory practice without any extended discussion of their possible threat to public health.

Nor did laboratory safety receive any unusual degree of attention when biologists turned en masse to the study of animal tumor viruses in response to the multimillion-dollar cancer crusade launched by the Nixon administration. Many of these researchers were biochemists or molecular biologists, untrained in the safety techniques that are second nature to the microbiologist, and some tended to regard tumor viruses as just another chemical reagent.

The often cavalier attitude toward safety was noted at the time, in 1973, by W. Emmett Barkley, the biological safety expert at the National Cancer Institute. "In the majority of labs we visit we see things that ought to be corrected," Barkley remarked. "The greatest offenders are university labs, not industrial labs. Most people working with tumor viruses have been exposed to some extent."[1] Robert Pollack, a cancer researcher then at Cold Spring Harbor Laboratory, described the situation as a "pre-Hiroshima condition—It would be a real disaster if one of the agents now being handled in research should in fact be a real human cancer agent."

Similar concerns were expressed at a special meeting on laboratory biohazards held in January 1973 at the Asilomar conference center, in Pacific Grove, California, site of the landmark meeting on gene splicing that was to take place two years later. Edwin Lennette of the California State Department of Public Health complained that the graduates he was teaching had an excellent appreciation of microbial genetics but little or no idea of how to protect themselves or innocent bystanders from pathogenic agents. "I am appalled at some of the techniques and procedures I see used by people who work with known or potential pathogens, but with little or no exposure to medical microbiology," Lennette observed, "and I think the time has come, if it is not long overdue, to remedy such situations."[2]

Discussion of the increasing number of scientists working with animal tumor viruses prompted one participant, James Watson, the co-discoverer of the double helix, to remark that the National Cancer Institute, the source of funding for many of the experiments, "avoids living up to its moral, if not legal, responsibility by declaring almost all the viruses we work with as unlikely to be of sufficient long-term danger" to require special facilities. Not the inconvenience of special facilities, but the "awfulness of the alternative possibility," should be the grounds for decision, Watson implied.[3]*

Safety consciousness among researchers is probably much keener now than it used to be. But one reason why safety was long neglected is the high-pressure atmosphere and intense rivalry of modern science, particularly in such fast-moving fields as molecular biology. The fierce pace of competition, though highly efficient at getting results, does not encourage researchers to handicap themselves with excessively rigorous safety precautions. Scientists prize their independence and do not lightly offer or receive unsolicited advice on such matters as how to run their laboratory. There is also a strong tradition of freely exchanging materials. Before the debate over gene splicing, any researcher who asked his colleagues for a guarantee that they would observe certain safety precautions would risk being accused of trying to gain an unfair advantage.

This is exactly what happened to Andrew Lewis when he made what was perhaps the first serious attempt to address the hazards of laboratory-created agents. Lewis is a virologist at the National Institute of Allergy and Infectious Diseases, part of the National Institutes of Health in Bethesda, Maryland. Working with the monkey virus known as SV40, he developed a virus that was a cross between SV40 and one of the adenoviruses that are the cause of the common cold. The hybrid had certain properties that made it particularly desirable as a research tool, and he soon began to receive the usual requests for samples from virologists in the United States and abroad.

But Lewis was concerned that the adenovirus part of the hybrid might enable it to infect the human nose and throat, exposing the victims to the unknown effects of the SV40 part of the hybrid. Since adenoviruses can infect large proportions of the population, and since SV40 is known to cause tumors in certain animals (though not, it seems, in man), there seemed to be reason for handling the viruses cautiously. With the advice and approval of the director of his institute, Lewis drew up a memorandum of understanding that asked those receiving the hybrid viruses to take special safety precautions and not to pass them on to anyone who refused to do likewise.

Even this seemingly innocuous restraint on the tradition of free exchange ran into hostile resistance from some of Lewis's colleagues. The directors of several well-known virology laboratories initially declined to sign the agreement, Lewis recalls, among them three of the group that was to play a leading role in the gene-splicing debate —Paul Berg of the Stanford University Medical School, Daniel Nathans of the Johns Hopkins University School of Medicine, and James Watson of the Cold Spring Harbor Laboratory.

Watson even tried to strong-arm Lewis into handing out his viruses in the usual no-strings manner. During the coffee break at a conference held in Cold Spring Harbor in 1971 Watson threatened the younger scientist with action from Congress, intervention by the director of the National Institutes of Health, and collective pressure from the scientific community if the viruses were not immediately made available to everyone.

"I just told him there were very few people who could respond to that kind of pressure," Lewis recalls. Though Lewis did not budge from his position, his faith in his colleagues' powers of voluntary self-restraint was somewhat shaken. When the National Academy of Sciences set up a committee in late 1973 to study the implications of gene splicing, with Berg, Nathans, and Watson among its members, Lewis wrote to the committee about his experiences with the hybrid viruses. Mentioning that he had been "confronted with threats [from one of the members of your committee]," he observed: "It is unlikely that in the competitive atmosphere in which science functions that broad unenforceable requests for voluntary restraint will contain the potentially hazardous replicating agents which arise from the widespread application of the plasmid recombinant technology."[4]

Lewis's attempt to ensure safe handling of his viruses aroused his colleagues' antagonism, but it may also have set some of them thinking about the issues he raised. It was in fact in Berg's laboratory that the conditions were created for the next flashpoint of the debate. Berg, like many other biologists, was working with SV40 and hoping to learn, among other things, which of the virus's three genes causes tumors and in what manner. He had developed a prototype version of the gene-splicing technique with which he proposed to insert the DNA of SV40 into *E. coli*. The chief difference between Berg's technique and that based on pSC101 was that Berg had to make his own sticky ends by a laborious system involving several different enzymes. It was not at that time known that restriction enzymes automatically leave sticky ends when they sever the double helix. That important fact was discovered later, in November 1972, by three other Stanford scientists: Janet Mertz, Ronald Davis, and Vittorio Sgaramella.

In the summer of 1971 Janet Mertz, then a graduate student of Berg's, was attending a course on tumor viruses at Cold Spring Harbor. One of the lectures was given by Robert Pollack, a cancer researcher who was particularly worried about the number of untrained people coming into the tumor virus field and the fact that viruses like SV40 were being handled in highly purified, and therefore infectious, form. When Mertz described the experiment that Berg proposed to conduct with SV40, Pollack was horrified and called Berg to protest. "I had a fit," he told Horace Judson, author of an article in *Harper's*.[5] "SV40 is a small animal tumor virus; in tissue cultures in the lab, SV40 also transforms individual human cells, making them look very like tumor cells. And bacteriophage lambda [Berg's proposed cloning vehicle] just naturally lives in *E. coli* and *E. coli* just naturally lives in people. She seemed to see it as a neat academic exercise. And I said, of all the stupid things, at least put it into a phage [a virus of bacteria], then, that doesn't grow in a bug that grows in your gut! Because what if the combination escapes from the lab: then you have SV40 replicating in step with the *E. coli* and a constant exposure of the cells in your gut to the DNA of SV40. Which is a route for the virus that never occurs in nature and therefore something you might not be prepared to fend off."

There can be few things more exasperating than to have someone raise remote-seeming safety hazards about an interesting experiment, especially one that could throw light on an important aspect of tumor viruses. But Berg recognized that he had not considered the possible hazardous consequences of what he was making. "At first it got my back up," he said later of Pollack's call. But after asking around his colleagues, he learned that they too had doubts about the experiment.

One whom he consulted was Leslie Orgel, a chemist at the Salk Institute who studies the origin of life. "Orgel pointed out that if our batting average in being correct [in predicting the outcome of such experiments] was twenty percent, we would be doing very well. Yet even if it rated ninety-nine percent, a one-percent uncertainty would present an intolerable risk. I found I could not persuade myself that there was zero risk," Berg recalls. He decided to postpone the experiment.

Other scientists at the time had equally strong doubts about the planned experiment, news of which had spread quite widely. "The Berg experiment scares the pants off a lot of people, including him," remarked NIH virologist Wallace Rowe. Another virologist, George Todaro of the National Cancer Institute, said the experiment "is one of those which I think just shouldn't be done."[6]

Berg's self-restraint was a significant step in the slow awakening of concern among the scientific community. But the pivotal event that initiated the gene-splicing debate took place in June 1973 at the annual Gordon Conference, a high-level research meeting held in New England schools left empty for the summer vacation. Boyer gave a lecture in which he described the pSC101 technique and the experiment in which the gene for penicillin resistance had been transferred to *E. coli.* His talk provoked a discussion about the implications of the technique, after which several younger scientists came to the cochairpersons of the session, Maxine Singer of the National Institutes of Health and Dieter Söll of Yale, with the suggestion that there should be consideration of the safety issues.

There was some resistance to the idea, but Singer and Söll supported it, and a special fifteen-minute session was arranged on the last day of the conference. Many participants had left, but a majority of the ninety who remained agreed to send a public letter suggesting that the National Academy of Sciences appoint a committee to study the possible health hazards of gene splicing to laboratory workers and the public.[7] The letter, published in the 21 September 1973 issue of *Science*, referred to the unpredictable nature of organisms that could be created with the new technique and suggested that though no hazard has been established, the potential hazard should be seriously considered.

When the National Academy received the Singer–Söll letter, an official called Singer to ask what the Academy should do. She said they should ask Berg. At the Academy's request, Berg convened a meeting held at the Massachusetts Institute of Technology in April the following year. Those present included Berg, Daniel Nathans, and James Watson; David Baltimore of MIT, who was later to receive a Nobel Prize for his codiscovery of an interesting enzyme associated with certain tumor viruses; Norton Zinder of Rockefeller University, who codiscovered an important property of bacterial viruses called transduction; Sherman Weissman of Yale; and Richard Roblin of the Harvard Medical School.

Numerous requests had by that time flowed into Stanford for the pSC101 plasmid, some from researchers planning experiments that were certainly no less hazardous than the one Berg had postponed. One scientist proposed, for example, to link the plasmid to herpes virus, a group of agents that cause sores and other diseases in man. Cohen was sending out his plasmid with a warning that in essence asked recipients not to use it for thoughtless experiments. But for how long would the warning be universally heeded?

The National Academy group that met at MIT had little difficulty in arriving at the obvious decision: to recommend that an international conference be convened to discuss the issue further. But it would take several months to arrange such a conference, and interest in the research was picking up so fast that many "bad molecules" might be produced in the interim.

So the group took a more unexpected action. Members decided to ask their colleagues to hold off from the more obviously hazardous or thoughtless kinds of experiment until the conference had met.

The request may seem a small step, but it was without major precedent in the history of science. Before the Second World War the Hungarian–American physicist Leo Szilard had tried ineffectually to prevent nuclear physicists from publishing results that could be of possible aid to the Nazi regime. In November 1969 a team of Harvard scientists—Jon Beckwith, James Shapiro, and Larry Eron—announced they had isolated a pure gene from a bacterium and took advantage of the occasion to warn of the dangers of government use of science. But the Academy group's call for a voluntary moratorium, even though it applied to only a few types of experiments, seems to be the first time that scientists have suggested restraint on research itself. Members of the Academy group were aware of no precedent for their action. "It is the first time in our field," said Berg, who served as chairman of the group. "It is also the first time anyone has had to stop and think about experiments in terms of the potential hazard."

Having made their decision, the Academy group asked some of the most active practitioners of gene splicing to join in their appeal for a moratorium. Those co-opted included Cohen and Boyer, two of the four-member team whose invention of gene splicing had been published just a few months earlier, in November 1973, together with David Hogness and Ronald Davis of Stanford. The group's appeal was presented in the form of a letter published in the 26 July 1974 issue of *Science* and *Nature*, weekly scientific journals edited from Washington and London, respectively, which have a wide circulation among research scientists in all the basic disciplines.

The letter began by observing that however fruitful experiments with the new technique might be, they "would also result in the creation of novel types of infectious DNA elements whose biological properties cannot be predicted in advance." Citing their "serious concern" that some of these artificial DNA molecules could prove hazardous, the signatories asked colleagues throughout the world to join them in deferring certain types of experiment until the international conference could decide how to proceed further.

Two types of experiment fell under the group's request for a moratorium:

- Putting the genes for antibiotic resistance or toxin formation into bacteria that do not at present possess such properties; and constructing plasmids with new combinations of antibiotic resistance

- Joining all or part of the genes of animal viruses onto a cloning vehicle, such as a plasmid or another virus.

In addition, the group asked that plans to put any genes from animal cells into bacteria should be "carefully weighed" in view of the possibility of releasing one of the tumor virus elements that many types of animal cells carry integrated within their own DNA. Such experiments, the group warned, "should not be undertaken lightly."

Only two categories of experiments, a minute fraction of the possibilities opened by gene splicing, came under the requested moratorium. Yet no one could be certain that the moratorium would be honored even in the United States, let alone worldwide. Scientists as a group, especially those at the leading edges of their field, are strong willed, highly motivated, and intensely competitive. It could not be taken for granted that they would rate the Academy group's opinions on safety matters any higher than their own.

"Caltech and Harvard will respect them, but those not in the elite will see no reason to hold off," commented the editor of a biological journal when the letter was published. "Anyone who wants to will go ahead and do it," was the gloomy prognosis of a scientist familiar with biohazard issues, who predicted that gene splicing would become "a high school biology project within a few years."

Members of the Academy group were more hopeful. "I think the recommendations will stick because they are reasonable, and the better part of the scientific community recognizes the need for caution. The worst part will be under a kind of moral pressure to go along with the majority," David Baltimore predicted. Berg too subscribed to the power of peer pressure: "Anybody who goes ahead willy-nilly will be under tremendous pressure to explain his actions."

As it turned out, the moratorium was scrupulously observed throughout its requested duration, from July 1974 until the convening of the conference seven months later. As far as is known, it was heeded by scientists in Europe and the Soviet Union as well as by those in the United States. In England the Medical Research Council (the equivalent of the National Institutes of Health) ordered its scientists not to undertake any of the experiments mentioned in the Academy group's letter.

Allegiance to the moratorium was doubtless encouraged by the fact, as Baltimore bluntly noted, that those suggesting it represented both the East and West Coast elites of American molecular biology. Moreover, their letter was a politically dexterous document, hewing to the middle of a range of options in moderate terms that disarmed opposition from both flanks. The group thus achieved its principal purpose of buying time until a legitimate forum of discussion could be convened, the international conference that was to take place at Asilomar in February 1975.

NOTES

1. This and all other quotations, except where otherwise noted, were made to the author in the course of interviews for articles in *Science*.

2. *Biohazards in Biological Research* (Cold Spring Harbor, N.Y.: Cold Spring Harbor Laboratory 1973), p. 353.

3. Ibid., p. 351.

4. Lewis to the Committee on Recombinant DNA, National Academy of Sciences, 29 November 1974.

5. Horace F. Judson, "Fearful of Science," *Harper's Magazine*, June 1975, p. 72.

6. "Microbiology: Hazardous Profession Faces New Uncertainties." *Science* 182 (9 November 1973): 566.

7. The most detailed description of these events is given by William Bennett and Joel Gurin in "Science That Frightens Scientists," *The Atlantic*, February 1977.

5

THE CONFERENCE
AT ASILOMAR

AT LEAST A footnote in the history of science will be reserved for the meeting that took place February 24–27, 1975, at the Asilomar conference center in Pacific Grove, California.[1] Asilomar is an abandoned chapel set in pine groves bordering the ocean. Its grounds are part of the wintering area for the western American population of monarch butterflies.

Not quite monarchs but at least the paladins of their own special world, the directors of research laboratories and arbiters of scientific fashion descended on Asilomar from all over the world—from England, Germany, and France; Russia, Japan, and Australia; Canada, Holland, Italy, Belgium, Sweden, and Denmark.

The ninety American and fifty foreign scientists had convened to discuss not the ethical or long-term implications of gene splicing but a specific practical matter: whether or not experiments with recombinant DNA presented any health hazard to researchers or the general public.

Drawing up safety regulations is not a matter of overwhelming concern to most scientists, but almost nobody invited to this particular conference failed to attend.

41

The chances of the conference ending in disagreement were probably rather better than even. Few scientists are accustomed to being told how to conduct their experiments, and no one acquainted with the near limitless powers of the new technique would lightly accept any serious restrictions on its use. Yet how could the public be informed that the moratorium was at an end when the unknown hazards that had caused it to be invoked in the first place were just as unknown as before?

A consensus was probably the last thing that the organizing committee of the conference expected to achieve. The organizing committee consisted of Paul Berg, David Baltimore, and Richard Roblin from the Academy group that had summoned the conference; Maxine Singer, coauthor of the Gordon Conference letter that first brought the issue to the scientific public's attention; and one new arrival to the debate, Sydney Brenner of the Laboratory of Molecular Biology in Cambridge, England.

Of all the members of the organizing committee, Brenner impressed his views most vividly upon the conference. A distinguished pioneer in deciphering the genetic code, Brenner is a speaker of entrancing lucidity, with a gift of such sharp wit that his audience takes its mere absence as a sign that he has something serious to say. Unlike some of the scientific leaders at the conference, whose style is to direct large teams of graduate students without ever touching a test tube, Brenner still enjoys doing experiments with his own hands. The public tends to think of scientists as men interested exclusively in facts and logic, which is true enough of the duller ones, but the best researchers, when they talk about their work, exude a sense of intuition about nature. One of the reasons why Brenner's views carried such weight at the Asilomar conference is that his remarks were always imbued with an inner feel for the way things ought to be.

Brenner's position was that whatever safety procedures were drawn up should be of such evident stringency that no one could reasonably accuse the scientific community of being self-serving. Many scientists favored devising safety regulations that would "minimize" the risk. Brenner's idea of a successful safety standard was one that would certainly require future revision, but revision downward because it was patently too strict, not upward because there had been an accident.

The organizing committee's chief adversaries were James Watson, and Joshua Lederberg of Stanford University, renowned for his pathfinding work on the genetics of bacteria in the 1930s and 1940s, when many doubted that bacteria had any genetics. Like a pair of *enfants terribles*, the two Nobel laureates were constantly discovering holes in the committee's positions and—although there seemed to be no concerted campaign—breaking the ice for the faction among the younger scientists who were eager to get research under way on the easiest terms available.

The conference was opened with a stern invocation from Baltimore. He reminded people that if they failed to reach consensus, if they split along lines that he could easily imagine, there was no one else to appeal to, and the conference would have failed in its duty.

How was the consensus to be determined? someone inquired. "The procedures by which the consensus will be determined will be largely determined by the extent of the consensus," was Baltimore's steel-in-velvet reply.

The first subject of discussion was *Escherichia coli*, the bacterium that inhabits the gut of man and other warm-blooded animals. The particular strain used in all research laboratories, designated *E. coli* K12, was isolated in Stanford in 1922, from the feces of a patient recovering from diphtheria. *E. coli* K12 is commonly believed to have become less robust than the natural strains of *E. coli* because of its long domestication in the test tube. Since K12 would serve as the host for many recombinant DNA experiments, whether or not it could infect man was a question of prime importance.

Two English microbiologists, Ephraim Anderson and H. Williams Smith, described the results of drinking bacterial cocktails—K12 in a glass of milk. Williams Smith used himself, Anderson had dosed volunteers, but both reported that K12 did not succeed in colonizing the gut and becoming part of the resident flora. Anderson, however, noted that the bacteria did linger long enough in the bowel to transfer an occasional plasmid to other bacteria.

One chain of events that might lead to a laboratory accident with gene splicing would be if an *E. coli* K12, containing a plasmid laden with harmful foreign genes, should infect a laboratory worker and either colonize his gut or transfer its plasmid to the *E. coli* bacteria already there. The Williams Smith and Anderson data suggested that such a scenario was rather unlikely yet not completely impossible.

The conference next heard from Harold Green, one of the few nonscientists invited to participate. A Washington lawyer interested in the interaction of science and public policy, Green's chief point was the warning that at the inception of any new technology the benefits, which seem near at hand, are often permitted to outweigh the risks, which appear more remote.

Two examples of this point that surfaced later in the meeting, though Green did not specifically cite them, are the biological technologies of polio vaccine and hybrid corn. Who could have advocated foregoing the benefits of polio vaccine in the 1950s? Yet the vaccine received by up to thirty million Americans and by probably a greater number of Russians is now known to have been contaminated with live SV40 virus from the monkey cells used to culture the vaccine.[2] SV40 causes tumors in hamsters and induces cancerlike behavior when it infects human cells cultured in the test tube. On present evidence, at least, it seems that the virus does not cause cancer or other disease in man. But for that stroke of fortune, the risks of the early polio vaccines would have far outweighed the benefits.

Another biological technology whose benefits and risks were incorrectly judged was that of hybrid corn. By 1970 a major portion of the nation's corn was derived from hybrids all of which had the same genetic trait. When a new strain of fungus eventually emerged which was well adapted to that trait, much of the crop was suddenly vulnerable. It was perhaps fortunate that only 20 percent of the national corn crop was wiped out.

The Asilomar conference's first attempt to grapple with its own risk-benefit problem was a document that ranked recombinant DNA experiments according to their expected degree of hazard and suggested a matching range of physical containment procedures that would reduce the hazard to an "acceptable" level of risk.

The working group that had prepared the document considered that it could serve as the basic solution to the problem. The proposal was attacked by Lederberg for being too precise; legislators would translate it into a "message from on high from which all further exegesis is forbidden." But the heavier onslaught came from the opposite quarter, from Brenner, who declared that the conference should not act as a licensing authority and that if it did, he would resign from the organizing committee.

The issue, Brenner continued, "is how to proceed in this area without presenting any risk to ourselves, to the innocent within our institutions, or to the innocent outside them. . . . I think there are people here who feel that there will be a negotiable set of compartments and that any particular compartment would comply with their local conditions. I am utterly opposed to that way of thinking. . . . If people think they are going to get a license from this meeting, a notice they can put up on their door, if they are just pretending there is a hazard and are going along with it just so that they can get tenure and be elected to the National Academy and other things that scientists are interested in doing, then the conference will utterly have failed."

Next Watson stood up and said he thought the moratorium should end. This was surprising because Watson had been a member of the Academy group that invoked it. The reason, Watson explained, was that "when we met, I thought we should have six months to see if we could hear anything that would frighten us. As someone in charge of a tumor virus laboratory, I feel we are working with something which is instinctively more dangerous than anything I have heard about here. . . . The dangers involved [in gene splicing] are probably no more dangerous than working in a hospital. You have to live with the fact that someone may sue you for a million dollars if you are careless. That sounds very negative and right wing, but I don't see any other way of doing it."

Watson was speaking for one side of an important divide in contemporary molecular biology, the cancer virologists and old-school microbiologists who are used to dealing with highly infectious agents and who could be recognized at the conference by their habit of using their elbows to shut off washroom faucets. Several of them spoke with horror of the "sloppiness" and "prostitution of microbiological technique" of the younger molecular biologists who have recently invaded the field. "It is the *E. coli* people who are screaming," remarked one tumor virologist who, like Watson, believed that the standard safety procedures for handling infectious agents were perfectly adequate for handling gene splicing as well—as long as the individual scientist took the procedures seriously.

Over the next two days the conference heard the reports of two other working groups. One group tried to assess the hazards of doing shotgun experiments with the DNA of various species. The other considered experiments involving animal viruses and concluded that the only safety precautions necessary were those for handling tumor viruses, as specified by the National Cancer Institute. One of this group's members disagreed: Andrew Lewis, perhaps remembering his own difficulties several years before in interesting his colleagues in safety issues, argued that no gene-splicing experiments with animal viruses should be undertaken until means of biological containment had been developed.

The idea of biological containment, of which Brenner was a principal proponent, is to disable bacteria genetically in such a way that they would be unable to produce certain essential chemicals. The crippled organism would be able to survive only as long as these chemicals were supplied in its laboratory diet. Should the organism infect a researcher or escape from the laboratory, it would rapidly die for lack of its special diet.

The scheme advanced by Brenner and others was to set up a system of biological containment—chiefly using crippled *E. coli* K12—alongside conventional means of physical containment, thus providing a double set of defenses against the risk of a microbial breakout.

The idea seemed both elegant and eminently practical—some delegates were even predicting that crippled bacteria could be devised within a matter of weeks. But for this belief, which was to prove hopelessly optimistic, the conference might have been far less ready to embrace biological containment as the central solution to the problem of hazard.

The three working groups at the Asilomar conference were largely composed of the researchers whose plans would be most affected by any safety regulations. For many of them the chief concern was that the conference agree to some workable set of safety prescriptions so that they could get back to their research. But attempts to have any specific set of prescriptions debated in detail were repeatedly sidetracked by participants who raised more general issues. "The consensus here is that people want guidelines and containment so they can go and do their experiments, but no one will come out and say it—they're all chicken," remarked the leader of one of the three working groups.

The dilemma of those who wanted clear guidelines was that, despite the various attempts to rank experiments in order of risk, no one had any real idea of what the risk might be or how to measure it, a point that emerged clearly in the following interchange:

Ole Maaloe *(University of Copenhagen):* I think we are misbehaving ourselves very considerably at this moment because it is nonsense to my mind to try to proofread your report as if it were a legal document. . . . We are trying to give the public some assurance that we are thinking seriously about what we are doing. To imagine that we can lay down even fairly simple general rules would be deceiving ourselves. . . .

Lederberg *(Stanford University):* Either we tear up this paper, or else we go into it very carefully. If it is likely to be crystallized into legislation, we had better be sure that it is right. . . .

Berg *(Stanford University):* We have to make some decisions. If you concede there is a graded set of risks, that is what you have to respond to.

Watson *(Cold Spring Harbor Laboratory):* But you can't measure the risk. So they want to put me out of business for something you can't measure.

Joe Sambrook *(Cold Spring Harbor Laboratory):* As far as I am concerned, there is no absolute containment and all containment is inefficient.

Roy Curtiss *(University of Alabama):* There is no way to decide right now what a safe organism is or what a safe cloning vehicle is.

Cohen *(Stanford University):* If the collected wisdom of this group doesn't result in recommendations, the recommendations may come from other groups which are less qualified.

Robert Sinsheimer *(Caltech):* I have sensed almost no willingness in this group to concede that there are some experiments which should not be done. At the same time Watson says, quite correctly, that there is no way to measure the risk. It would seem to me that in the end we will be regulated. We would be in a better position to face that if we take the position that some of the higher [risk] categories of experiment should not be done at all until more information is available. I cannot think of anything that would impede science more than if there was an epidemic around Stanford.

The final evening of the conference was given over to a panel of lawyers, whose reflections were not wholly encouraging to their scientific colleagues. Laboratory workers injured by the experiments could sue, the lawyers observed. "Many people have talked about this research under the banner of academic freedom," observed Alexander Capron of the University of Pennsylvania Law School. But, he added, "by overstating their case they risk provoking greater restriction. Freedom of thought does not encompass freedom to cause physical injury to others, and 'prior restraint,' as one person here has termed it, is an absolutely appropriate response where irreversible harm is threatened."

Perhaps the most telling point was made by Roger Dworkin of the University of Indiana. Scientists, in his opinion, no longer enjoyed their traditional immunity from being called to account. "You people," the lawyer observed, "being experts, have a tremendous amount of power. Even though the general reservoir of respect for science is less than it was, many people are still mesmerized by science. If this body proposes any kind of reasonable scheme at all for dealing with biohazards, I think the scheme will receive tremendously high respect and will have high odds of adoption. Because the law has a tradition of allowing expert groups to regulate themselves.

"But there have been disasters when groups have not used that discretion wisely. Take medical malpractice. The doctors abused their privileges: they are being massacred in the courts. Any appearance of self serving will sacrifice the reservoir of respect that scientists have and will bring disaster upon them."

Dworkin's observations were coldly received, but Brenner then developed the same theme. "The issue that I believe is central," he said, "is a political issue. It is this: We live at a time when I think there is a great antiscience attitude developing in society, well developed in some societies, and developing in government, and this is something we have to take into consideration. . . . Who really believes that natural science will increase your GNP? Maybe this is the end of this era. It is very hard to tell in history where you really are. . . . I think people have got to realize there is no easy way out of this situation: we have not only to say we are going to act, but we must be seen to be acting."

With such thoughts possibly in mind, the delegates assembled the next morning to learn the organizing committee's opinion of what their consensus was. The document the committee had finished at about 4:30 A.M. that morning possessed considerably more bite than any of the schemes produced by the three working groups.

The most innocuous category of gene-splicing experiments was to take place under the "low risk" conditions of physical containment recommended by the National Cancer Institute for handling tumor viruses such as SV40. All others were to be done in "moderate risk" or "high risk" conditions and were, moreover, to await the development of crippled bacteria. High-risk physical containment involves laboratories with negative air pressure, special safety cabinets, and shower rooms to be used on leaving; these are the conditions used for handling the most dangerous known human pathogens, such as Lassa fever and bubonic plague.

Since the crippled bacteria were not expected to be available for a matter of several months—Cohen at one point likened their arrival to waiting for the Messiah—the effect of the organizing committee's document would clearly be not only to continue the moratorium but to extend it to many kinds of experiments that had previously been excluded.

Berg, as chairman, presented the document with the observation that it was the organizing committee's opinion of the consensus of the conference. This was somewhat adventurous in that, as one participant promptly pointed out, some of the document's more rigorous prohibitions were being introduced for the first time. Several members of the working groups began to criticize certain features of the document for being excessively strict.

It soon became apparent that the organizing committee did not intend to have its statement amended from the floor and was also reluctant to test the actuality of the "consensus" by any measure so crude as a democratic vote. But a vote was forced, and to the committee's evident surprise the document turned out to be reflecting a consensus that was every bit as solid as the committee was pretending. The conditions specified for all three classes of experiment were adopted almost unanimously, with at most five hands raised in opposition.

The committee lost on only one point: a motion that the most dangerous types of experiment, such as inserting the genes for botulinus toxin into *E. coli,* should not be performed under any circumstances. After some initial resistance from the committee, the motion passed with only five dissenting votes.

With the business of the conference essentially complete, a spokesman for the Russian delegates stood up and said that the organizing committee's statement was reasonable and acceptable and would be a useful guide for the relevant discussions in the Soviet Union.

A final vote was taken to approve the entire document, which passed with only Cohen and Lederberg dissenting. One of the lawyers, Daniel Singer, said it had been a moving experience for him as an outsider to watch the group grappling with a very difficult problem. Baltimore then paid tribute to Berg for his role in organizing the conference, and with that the meeting was over.

At a press conference held afterward, Berg, weary from lack of sleep, was asked if he thought the Academy group's original call for a moratorium on certain experiments had been an overreaction. "Not at all. It has raised the level of discussion about this issue. Six months ago we had daily phone calls asking for pSC101. I would ask people what they wanted to do with it. Some of them had horror experiments planned with no thought of the consequences. But I was in the same position myself because I was going to do a similar experiment two years ago and someone called me up and asked if I had thought of the consequences."

There was a direct line of descent from Pollack's phone call, to Berg's scruples, to the Gordon Conference discussion and the Singer-Söll letter, and from there to the Asilomar conference. Yet the sequence of events was by no means a foregone conclusion, and it was a certain achievement of the international scientific community that it was able to subordinate its interest in unfettered research to agreement on a set of general principles for ensuring the safety of the new experiments. But a more divisive decision was yet to come; the translation of the principles into a detailed code of practice.

NOTES

1. This chapter is adapted from my article in *Science* 187 (14 March 1975): 931.

2. Keerti Shah and Neal Nathanson, "Human Exposure to SV40: Review and Comment" (Resource document prepared for the Asilomar conference, Pacific Grove, Calif., February 1975).

6

ASSESSING THE HAZARDS

THE MEETING AT Asilomar served as an international constitutional conference on the governance of gene splicing. Having agreed on the outline of the constitution, each nation represented there was free to set about writing its own laws within that general framework.

In practice, other countries waited and watched to see what the United States would do, and the rules they produced were very similar in substance, though not in form, to those devised by the National Institutes of Health.

Because of their worldwide influence the NIH rules, and the manner in which they were drawn up, are matters of some relevance to the debut of the gene splicer's art.

The unanimity of the Asilomar conference was no accurate guide to the dissensions that followed, or to the surprising difficulty of framing a solution acceptable to both scientists and the public. The details of two-year debate that ensued between the Asilomar conference and the intervention by the U.S. Congress in spring 1977 are perhaps best understood in the context of the general course the debate followed.

The National Institutes of Health set up a committee

of research biologists to translate the Asilomar principles into a specific code of safety conditions for each conceivable kind of experiment. The committee held its first meeting the very day after the conference—on February 28, 1975—yet its safety guidelines did not appear in final form until sixteen months later, on June 23, 1976.

The reason for the delay was the flurry of protest touched off in the scientific community by the committee's first draft. The safety conditions proposed appeared so lenient to other biologists that the NIH sent its committee back to the drawing board.

The final version of the guidelines was more stringent, but just as it was being readied for publication, the committee's general approach came under criticism from two different fronts.

First, the committee was criticized on matters of principle by two scientists of considerable eminence, Erwin Chargaff of Columbia University and Robert Sinsheimer of Caltech. Chargaff derided the committee's two safety systems as a "smokescreen" and a "cardinal folly" and accused his colleagues of overlooking the ethical implications of the research. Sinsheimer criticized the committee for having focused so narrowly on the issue of health hazards that it had ignored the evolutionary consequences of gene splicing.

The emergence of these two formidable critics fractured the consensus that had begun to build up around the NIH guidelines and demonstrated for the first time that there were serious divisions within the scientific community. Although Sinsheimer, at least, believed that the health hazard issue had been adequately taken care of in the guidelines, his and Chargaff's criticisms made it easier for other scientist-critics to dispute the committee's approach on health hazards as well as other issues.

Once under criticism, the NIH committee, regardless of the merits of its case, in addition found itself vulnerable to the charge of conflict of interest, since almost all its members were biologists who might some day wish to use the technique, and three were already doing so. Because most of the decisions about appropriate safety levels involved questions of judgment rather than hard scientific fact, the charge of unwitting preference—no one has accused the committee of deliberate bias—was hard to rebut.

Criticism from within the scientific community was followed by a much less well-mannered attack from outside. Scientists at Harvard had decided to conduct certain gene-splicing experiments in an old, ant-infested building, thereby handing the university's long-time adversary, Mayor Alfred Vellucci of Cambridge, a ready-made pretext for clangorous intervention. On the same day that the NIH published its guidelines Vellucci held the first of two widely publicized hearings on recombinant DNA research at Harvard and MIT.

Vellucci's threat to ban the research evaporated eight months later when his own city council voted to let it continue. The more significant effect of his actions was to stir up public concern about gene splicing. Cities from San Diego to Bloomington, Indiana, began to take a look at what was happening on their local campuses. State authorities in both New York and California prepared legislation to govern recombinant DNA research.

The NIH guidelines has the serious disadvantage of not applying to industry, which was taking a particular interest in the technique. The likelihood that local authorities would write their own laws, leading to a crazy-quilt pattern of differing standards across the country, made federal legislation inevitable. In spring 1977 Congress decided to act on the issue, marking the end of the first phase of the debate.

By that time so many words had been written, hearings held, reports issued, rumors spread, and insults hurled that the thread of the debate had become almost hopelessly tangled, the only evident pattern being that much of it was wound in circles.

The dominant theme of discussion, both during the formulation of the guidelines and thereafter, has been the health hazard issue. But a few interesting tangents have been projected, some of which may prove of more lasting importance. One is the emergence of Sinsheimer as the most thoughtful and generally respected critic of the NIH committee's approach. On a wide spectrum of issues, ranging from the evolutionary impact of recombinant DNA research to the capacity of social institutions to deal with its applications, he has provided a sustained flow of original and measured analysis.

A major theme developed by the critics is that not all new knowledge is necessarily beneficent and that gene splicing is only the first step toward human genetic engineering. The proponents have countered that all technologies can be controlled so as to maximize benefits and minimize risks and that history shows the dangers of restricting freedom of inquiry. As the public began to take an interest in the debate, the proponents have stressed the practical benefits that may be expected to flow from the research.

The division of the scientific community into proponents and critics of the NIH approach is a somewhat rough categorization; there is a wide spectrum of opinion within each camp. But the division exists and is a far cry from the unanimity that prevailed at Asilomar. Before tracing the development of the division, it is convenient to leap ahead from Asilomar to the focus of the division, the approach embodied in the final version of the NIH guidelines.

The purpose of the guidelines is to minimize the health hazards of recombinant DNA research. What kinds of hazards were perceived and what was the NIH committee's strategy for dealing with them?

⌐The possible hazards of gene splicing all stem from the fact that there is as yet no predictive theory of evolution, no way of forecasting the effect of transferring genes from one species to another. This state of ignorance is not surprising: not until the invention of gene splicing was it even suspected that genes could be transferred to and survive in a strange organism.⌐

To assess the real hazards of gene splicing it would be helpful to have copious information in a range of subjects—such as the ecology of *E. coli* and the natural rates at which genes are transferred in nature—that are at present largely unknown territory because they have never before seemed important.

In the absence of the required information it has been necessary to guess what kind of hazards might arise and how best to forestall them. As shown by the case of nuclear safety, where engineers thought of everything that might wreck a nuclear power plant except a man inspecting with a candle—that was the fate of the TVA's plant in Browns Ferry—it is the hazards a person is not clever enough to think of that are most likely to hit him from behind.

But whereas nuclear technology is completely unforgiving of human error, with recombinant DNA the odds should be stacked the other way. Adding foreign genes to an organism is probably comparable to a mutation, and the great majority of mutations are either neutral or a handicap to the organism in which they occur. So the apprehension about gene splicing is reduced to the rare occasions on which foreign genes may offer some usable new capability to their host, enabling it to exploit a new niche and present some degree of threat to man or the environment.

Such a possibility, however unlikely in theory, cannot be dismissed out of hand. In fact the first gene-splicing experiment ever performed was of this general nature. It involved transferring the gene for resistance to penicillin from one bacterium to another. In this particular case, the wild form of the recipient bacterium already possessed the gene, but there is little doubt that transfer of resistance genes to pathogenic bacteria which have not yet acquired them could make them more dangerous to man.

The following types of hazards—or "disaster scenarios"—are among those raised and discussed by scientists during the formulation of the NIH committee's guidelines. The interest of the scenarios is not so much that any is likely to occur—the opposite is the case—but that they influenced the NIH committee's thinking in devising its guidelines.

The most obvious class of hazards arises from the ubiquitous use of *E. coli* as the standard host for foreign genes. And the worst conceivable hazard is that man's normally peaceful microbial guest should suddenly be turned into an organism capable of causing epidemic disease.

How might this come about? A researcher, whether by design or accident, inserts into *E. coli* foreign genes that somehow convert it into a pathogenic organism. The virulent *E. coli* escapes from the laboratory by infecting a researcher or technician, who in turn infects his or her family and others, starting off the spread of the bacterium throughout the population. Establishing colonies in people's guts, just as does any other strain of *E. coli,* the escaped bacterium produces quantities of the protein specified by the foreign gene. The protein crosses the gut wall, exposing the individual to significant daily doses of the foreign protein.

The nature of the disease that might be produced would depend on the particular genes inserted into the bacterium. Several possibilities were discussed during the framing of the NIH guidelines.

One is the situation presented by the original Berg experiment, the insertion of SV40 genes into *E. coli.*

SV40 is a tumor virus, although its effects seem to be confined to lower animals. To take the worst-case analysis, however, it is assumed that the tumor-causing propensity of SV40 is somehow brought into effect in the individuals infected by the bacterium, setting off an epidemic of communicable cancer which is only discovered after however many years is the incubation period of the disease.

Another but related hazard that concerned the committee was the risk of unleashing a tumor virus in a shotgun experiment. Several species of animal are known to possess in their gene set sequences of DNA which match those of certain tumor viruses. In a shotgun experiment, which involves fragmenting and inserting the whole gene set of an organism into a population of *E. coli,* one of the resulting clones of bacteria might contain the DNA sequence of the tumor virus. Instead of being quiescent as before, when it was an integral part of the organism's gene set, the tumor virus sequence might somehow be activated, or "derepressed," by the shotgun procedure. The researcher might not suspect the presence of a tumor virus and even if he did, would not know which of the several thousand clones contained it. Bacteria from the tumor virus clone, as the disasterscape has it, would escape and cause disease by the usual route.

Shotgun experiments pose other conceivable hazards. Depending on the organism being shotgunned, one of its genes might specify a hormone or enzyme sufficiently similar to the human variety to interfere in the body's metabolism. Natural hormone and enzyme systems are governed by elaborate regulatory mechanisms; a source of hormone or enzyme located in the gut would not be subject to control. People infected with a gutful of bacteria producing the hormones which control reproduction, say, might find their own hormonal balances upset, and the epidemic spread of such bacteria might have certain demographic consequences.

Similar hazards could be raised by bacteria designed

to produce human insulin, human growth hormone, and other proteins of medical importance should they escape from the place of manufacture.

Another possible hazard raised by shotgun experiments concerns the production of proteins which are sufficiently foreign to trigger off the individual's immune response system into producing antibodies against them, yet which are similar enough to the human version that the antibodies attack the body's own protein as well as the foreign one. Attack by one's own antibodies is thought to be the basis of certain autoimmune diseases. One example concerns a protein involved in nerve-muscle communication known as the ACh-receptor protein. Rabbits injected with ACh-receptors from eels develop antibodies which attack both the injected ACh-receptors and the rabbits' own, with the result that the animals die of paralysis after two weeks. Thus a theoretical hazard of shotgun experiments with animals likely to produce proteins similar to man's is the danger of provoking immunosensitivity.

A final hazard associated with shotgun experiments is that the organism being shotgunned may be contaminated with other organisms, such as viruses, bacteria, fungi, or parasites, some of which may be capable of causing disease in their host or in other species. The genes of all these unintended passengers may end up in clones along with the organism's own genes.

Another *E. coli*–based scenario discussed by the NIH committee envisages the bacterium being endowed with the gene for cellulase, the enzyme that breaks down the plant structural material cellulose. Cellulose is digested in ruminants but not by man, and in fact undigested cellulose has the important role of giving bulk to the feces. By destroying the roughage in the gut, a cellulase-containing *E. coli* might produce chronic diarrhea, a condition that can be fatal if prolonged. Just such an organism, it turned out later, had already been constructed in 1975 by A. Chakrabarty, of the General Electric Research and Development Center, as part of a project to produce methane gas from sewage sludge. The organism was not created by the restriction enzyme method, and its manufacture would not be prohibited by the current NIH guidelines. After Chakrabarty had transferred the cellulase gene from another bacterium into *E. coli,* it occurred to him that the bacterium, should it infect someone, might not only cause diarrhea but would produce breakdown products which other bacteria in the gut might turn to gas. He therefore destroyed the organism.

A hazard of a different sort might arise from inserting bacterial viruses of unknown properties into *E. coli.* For example, botulinus toxin, one of the most lethal poisons known, is produced by particular bacteria, but the genes for the toxin are carried by a virus from one strain of botulinus bacteria to another. (Other toxins, such as diphtheria and streptococcus toxins, are also under virus control.) The botulinus bacteria are anaerobic, meaning they cannot grow in the presence of oxygen, but *E. coli* is not. "Putting [bacterial virus] genes into *E. coli* could well create a toxigenic *E. coli* that is not anaerobic, that doesn't need just badly canned beans to get to us. It could be all over," NIH committee member Wallace Rowe has observed.[1] Such a risk, though "very real" in Rowe's opinion, is also very minor in that of all the possible DNA sequences that can be put into *E. coli,* only a minute fraction would actually be dangerous.

The NIH committee gave most of its attention to the risks presented to man, but there is also the environment to consider. *E. coli* inhabits other warm-blooded animals and can survive in soil and water. There is a steady flow of genetic information from one bacterial species to another by bacterial mating and by the exchange of plasmids. (The worldwide spread of antibiotic resistance among bacteria is mediated by genes carried on plasmids.) *E. coli* exchanges genetic information with over thirty groups of bacterial species. Might not the spliced genes in a bacterium escape from the laboratory and enter into this flow of genetic information so as to produce some small but ultimately injurious change? Even the displacement of a microorganism at the bottom of a food chain might interfere with some ecological system of importance to mankind. "These types of subtle effects are difficult to predict and give me cause for concern about the inadvertent escape and survival of recombinant DNA molecules," NIH committee member Roy Curtiss has written.[2]

Most of these hazard scenarios, particularly those in which *E. coli* is converted to an agent of epidemic disease, cannot in any way be called probable. In order for them to occur, a series of independent events must take place, each one of which is in itself fairly unlikely. Just as the chance of a tossed penny turning up heads three times in a row is $\frac{1}{2} \times \frac{1}{2} \times \frac{1}{2} = \frac{1}{8}$, so the chance of a whole chain of events coming to pass is the multiple of the separate probabilities of each. A long chain with even a few improbable links has a notably bleak chance of being completed.

For *E. coli* to cause an epidemic it must first have harmful genes inserted into it; second, manage to infect and colonize a laboratory worker; third, be able to synthesize the harmful protein specified by the inserted DNA; fourth, release the protein in sufficient quantities that it can survive digestion and enter the bloodstream in sufficient strength to produce a malign effect; and fifth, be spread from one person to another rapidly enough to cause an epidemic.

Each of the links in this chain is an event that is possible but seems unlikely, at least on present evidence. Most disease-causing bacteria seem to depend not on one but on several genes for their pathogenicity, so that the chances of creating a conventional pathogen out of *E. coli* by insertion of a single length of DNA seem quite small. Infection and colonization, the second link in the chain, is a risk that the NIH guidelines are specifically designed to reduce by requiring physical containment and the use of K12, the laboratory strain of *E. coli.*

The third link is for the product of the gene to be produced by the host bacteria, a process known as expression. At the time the guidelines were being prepared, the general opinion on the NIH committee was that genes from eucaryotes, the cells of higher animals, would not be correctly expressed or even expressed at all in procaryotes such as bacteria. Thus the third link seemed hardly a link at all. But it now seems that this guess may be wrong, and that eucaryotic DNA may be expressed quite easily in bacteria.

The fourth link requires the harmful protein to cross the lining of the intestine and enter the bloodstream. The likelihood of such an event cannot be estimated because little is known about what makes a substance transportable across the gut wall. Proteins like insulin transport very poorly (which is why diabetics have to inject insulin instead of swallowing it). On the other hand, much larger proteins, such as botulinus toxin, seem to get across rather easily.

The possibility of epidemic spread, the final link, is also difficult to assess, particularly since the factors governing the natural spread and establishment of *E. coli* strains in the human population are not yet well understood. The K12 strain, however, mandated by the NIH guidelines for gene-splicing experiments, is probably even less infective than ordinary strains seem to be. Even if an epidemic were to start, the public hygiene measures that are successful against other gut-dwelling microbes, such as cholera and typhoid fever, presumably would also be effective against a rampaging *E. coli.*

Hence the chance that a careless or unwitting gene

splicer could set off an epidemic disease seems less than overwhelming. Nonetheless, gene splicing involves large uncertainties, and in uncertainty lies risk. It is this risk that the NIH committee's guidelines are designed to address.

NOTES

1. U.S. Department of Health, Education, and Welfare, *Recombinant DNA Research,* no. (NIH) 76-1138, 1 (National Institutes of Health, August 1976): 308.

2. Roy Curtiss, "Genetic Manipulations of Microorganisms: Potential Benefits and Biohazards," *Annual Review of Microbiology* 1976.

7
THE STRATEGY OF CONTAINMENT

BETWEEN HUMANS and the microbiological underworld there is an intimate and relentless intercourse. Humans shed bacteria and viruses from their skin, eyes, nose, and other orifices in a ceaseless torrent, trailing clouds of microbes as they go. Up to 60,000 bacteria a minute are released from the body on flakes of skin. Some flakes are small enough to penetrate the interstices of clothing. Others are wafted out from under the clothing as the occupant walks about, the clothes serving as a bellows to disperse the microbe-laden rafts into the surrounding air. (The relentless flood can be significantly stemmed, the NIH committee advises, "with the wearing of close-fitting and closely woven underpants beneath the usual laboratory clothing"—a counsel likely to be honored more in the breach than in the observance.) Bacteria stream forth from any kind of scratch, boil, or skin infection, and they are sprayed out as passengers on the cloud of airborne droplets generated by coughs and sneezes. A single sneeze may launch a million droplets. Even the act of conversation spreads bacteria. About 250 droplets, it has been estimated, are dispersed into the air for every 100 words spoken. The number of microorganisms carried by each droplet varies from person to person.

The goal of laboratory safety is to try to maintain a certain degree of separation between the microbes in the experiment and those in the experimenter. Only the most elaborate equipment will enforce complete separation between the two traffic flows, but in most cases the occasional interchange is not too serious. With highly contagious diseases, such as Q fever, as few as ten organisms constitute an infectious dose, but with *E. coli* perhaps thousands or millions of bacteria are required (the real number is not known) to establish a successful colonization or infection of the individual.

How can man and his experimental microbes be kept apart? For coping with the risks of gene splicing the NIH committee devised a two-fold strategy based on prohibition and containment.[1] Experiments that present a clearly identifiable risk are banned; those where risks can be conceived of are placed under a double system of physical and biological containment.

Six kinds of experiment are forbidden for the time being: (*1*) cloning any genes taken from disease-causing organisms classified as either moderate or dangerous pathogens; (*2*) cloning any genes from possibly harmful tumor viruses; (*3*) doing any gene-splicing work with the genes for potent toxins, such as botulinus toxin or snake venom; (*4*) doing anything with plant pathogens that might increase their virulence or range of target organisms; (*5*) transferring genes for drug resistance to any species of bacterium not known already to possess the genes; and (*6*) deliberately releasing into the environment any organism containing a recombinant DNA molecule.

The last prohibition refers to the projects for moving useful genes, such as those for nitrogen fixation, into crop plants or custom-designing organisms to replace pesticides. Such projects may be studied in the laboratory or greenhouse but cannot yet be tested outside it.

All other gene-splicing experiments are assigned to particular levels of physical and biological containment. The physical levels are designated P1 to P4, and the biological levels EK1 to EK3, the EK referring to the K12 strain of *Escherichia coli*.

P1 is a standard microbiological laboratory with no special equipment or safety procedures. The work is done on open benches; eating and drinking in the lab are discouraged but not prohibited, and the same is true of mouth-pipetting, a common laboratory procedure. (A pipette is a glass straw with a bulge in the middle into which an exactly measured volume of fluid can be drawn. Sucking too hard on the pipette and drawing into the mouth a few drops of liquid containing bacteria or viruses is a common cause of laboratory infections.)

P2, the next level up, is not significantly different from P1 except that operations such as blending, which produce considerable quantities of airborne droplets, or aerosols, must be conducted inside a safety cabinet. Eating and drinking in the work area are prohibited, as is mouth-pipetting—a mechanical pipetting device must be used instead.

P3 is the first level at which the laboratory requires special design and equipment. All work with organisms containing recombinant DNA molecules, or which would produce aerosols of the molecules, must be performed within biological safety cabinets. The cabinets are generally open-fronted boxes with a curtain of air designed to flow across the entrance so that no airborne particles or droplets leave the enclosure. Access to the lab is through a double door, and the air pressure is controlled so that air flow through the doors is only inward.

P4 is the level of physical containment used at the army's biological warfare laboratories and elsewhere to contain organisms that are of extreme hazard to man, such as the virus which causes Lassa fever. Entry and exit to the lab are tightly controlled: supplies come through air locks, wastes leave through double-doored ovens in which they are first sterilized, and people come through changing rooms in which lab clothes are donned on entering and showers taken on leaving. All lab work takes place in closed, airtight cabinets, in which the researcher works by putting his arms through portholes to which arm-length rubber gloves are attached.

Except at the P4 level, the cost of physical containment is not exorbitant compared with other research costs. A biological safety cabinet, for example, costs about $5,000. To convert an ordinary laboratory to P3 standards costs around $50,000, and rather than trying to convert an old laboratory into a P4 facility, it would be cheaper to build one from scratch, at a cost of at least $200,000.

The three levels of biological containment for recombinant DNA research are EK1, EK2, and EK3, defined as follows:

EK1 requires simply that the researcher use as the host for recombinant DNA molecules the standard laboratory strain of *E. coli*, known as *E. coli* K12. As his cloning vehicle he must use a plasmid—such as pSC101—that cannot be transferred to another bacterium during mating.* The purpose of this requirement is to reduce the risk that if the *E. coli* K12 should get into someone's gut, it might transfer its plasmid and inserted genes to other bacteria.

---

*A researcher who broke this requirement was told by the NIH committee that he would have to destroy his painstakingly constructed bank of some fifty thousand clones containing yeast genes and ten thousand clones of fruitfly genes. The bank was made in the interim period between the Asilomar conference and the publication of the NIH guidelines, and thus did not infringe any existing regulations. The researcher, John Carbon of the University of California, Santa Barbara, did not in fact use a transmissible plasmid as the cloning vehicle, but his *E. coli* already contained a second plasmid that mobilizes or makes other plasmids transferrable.[2]

EK2 refers to host–cloning vehicle pairs in which, should the bacterium escape from the laboratory, the cloning vehicle has only 1 chance in 100 million of surviving, as determined by laboratory tests. The basis of the reduced survivability is a genetically disabled strain of *E. coli* K12 that requires special laboratory chemicals to stay alive. If it escapes, the theory goes, it should quickly expire for lack of its special diet, and its gene-carrying plasmid should die with it. The only disabled bacterium so far developed is one named chi-1776 by its inventor, Roy Curtiss of the University of Alabama (chi for the University of Chicago, where Curtiss used to work, and 1776 as a gesture to the national bicentennial).

EK3 is an EK2 system whose lack of survivability has been demonstrated not just by laboratory tests (the requirement for EK2) but by actual feeding tests in man and other animals. Testing that an organism has only 1 chance in 100 million of surviving is technically arduous, and Curtiss's chi-1776 was only accepted by the NIH as an EK2 system in December 1976. By mid 1977 the bacterium had not yet passed its tests as a true EK3 system.

The NIH guidelines require gene-splicing experiments to be done under various levels of physical and biological containment according to their expected degree of hazard. Shotgun experiments, in which the whole gene set of an organism is inserted into clones, vary in containment level according to the evolutionary closeness of the species to man. Thus shotgun experiments with primates (monkeys, apes, and man) must be performed either under P3 physical containment and EK3 biological containment (P3 + EK3) or under P4 + EK2.

Shotgun experiments with other species of mammals may be done in P3 + EK2 conditions. With cold-blooded vertebrates, such as frogs, a favorite laboratory organism, containment goes down to P2 + EK2. Other cold-blooded animals, such as fruitflies, another laboratory workhorse, can be shotgunned in P2 + EK1 conditions unless they produce toxins, in which case higher conditions apply.

Plants can be shotgunned in P2 + EK1 unless they are pathogenic, and bacteria that habitually exchange genetic information with *E. coli* may be shotgunned in P1 + EK1, or P2 + EK1 if pathogenic. Animal viruses must be cloned in P4 + EK2 conditions, plant viruses in P3 + EK1, and bacterial viruses and plasmids under the levels appropriate for the bacterial hosts (generally forbidden altogether if the bacteria are moderately or highly pathogenic species.)

The guidelines have built into them several opportunities for reducing the required levels of containment. Primate shotgun experiments, for example, might seem to be effectively banned by the requirement of P4 + EK2 or P3 + EK3, since P4 laboratories are few and far between and EK3 systems do not yet exist. However, if the cells to be shotgunned are taken from germ-line tissue (eggs or sperm), which are less likely than other cells to be infected with viruses, the shotgun may be performed in P3 + EK2 conditions, which are more readily attainable.

A similar reduction is available for shotgunning frogs; if the cells are taken from germ-line tissue, the experiment may be done in P2 + EK1 instead of P2 + EK2.

All the above conditions apply only to the initial shotgun experiment. Once the researcher starts working with a particular clone, he may do his experiments under almost standard laboratory conditions (P2 or P1 + EK1) if he can show that the clone is pure and contains no harmful genes.

Adherence to all the rules laid down in the guidelines is enforced by a two-tier review system. Every university or institution where gene splicing is done has to set up a local biohazards committee, which certifies to the NIH that its researchers possess the appropriate physical containment required for their experiments. Before funding the experiments the NIH checks that the researcher has chosen the right levels of containment and that any downward revision—for example, on the grounds that a clone has been purified—is justified by the evidence.

How well will the guidelines work? It is important to note that the system is not designed to provide absolute containment. Some bacteria and viruses are bound to escape from even the most highly contained laboratory. The intention of the NIH guidelines is to reduce the traffic of escaping organisms to a level that will present no threat to man, beast, or the environment.

On the other hand, though absolute containment is not an objective, the guidelines are intended to be more than merely adequate to contain the risk; they are designed to provide a large margin of safety, so that any future revision will be toward laxer, not stricter, levels of containment.

The advantage of requiring both physical and biological containment is that each system serves as a backup if the other fails. But each system has its weak points.

The weakest point of physical containment is undoubtedly not the equipment but its human operators. Slobs can get themselves infected in P4 labs, whereas with suitable training and motivation a serious microbiologist can handle all but the most dangerous human pathogens in P1 or P2 conditions without getting infected.

The contribution to safety of any particular degree of physical equipment is probably small compared with that of the researcher who operates it. In fact, W. Emmett Barkley, the NIH's biological safety expert, has estimated that 90 percent of safety comes from good technique on the part of the investigator, 8 percent from the physical equipment, and 2 percent from the design of the facility.[3]

The current strength of the psychological factor is impossible to estimate. Many researchers involved in gene splicing are biochemists or geneticists who have not always acquired the techniques that are second nature to a trained microbiologist. On the other hand, the extensive debate about gene splicing has probably sensitized all researchers to the possible hazards of the technique. Then again, there is considerably more incentive for the researcher to take care if he is handling an organism he knows could be lethal to him than if he believes, as the chief drafter of the NIH guidelines once remarked, that the risks involved in shotgun experiments are "similar to being struck by lightning." Even researchers working with tumor viruses do not always pay scrupulous attention to safety regulations: a recent survey among workers at the MIT Biology Department and Cancer Center showed that "more than 80 percent ate or smoked or saw such incidents in areas where it was prohibited."[4] With gene splicing, it could be supposed, attention to safety will be scrupulous to begin with but may gradually slacken off in the absence of some definite reinforcement, such as an accident.

As for the physical equipment itself, it works well but not perfectly. Some five thousand cases of laboratory-acquired infection have been recorded over the past thirty years, but most are now believed to have occurred in laboratories without special containment facilities.[5] P3-level equipment offers significantly greater protection to the researcher than does that of P1, and P4 gives an improvement of similar magnitude over P3. The rate of infection at the Center for Disease Control was halved after the installation of P3-type biological safety cabinets in 1960.[6]

The most instructive information on the value of physical equipment in preventing infection was gathered at the U.S. Army's Biological Warfare Laboratories at Fort Detrick, Maryland. From 1943 to the closure of the laboratory in 1969 there were 423 cases of infection and 3 deaths. According to a report by A. G. Wedum, the laboratory's safety director, most of the infections occurred before the installation of biological safety cabinets in 1950, after which the accident rate improved considerably. [7]

The laboratory's researchers were handling some of the most infectious known organisms, such as the agents of anthrax, tularemia, and Q fever, yet in the last ten years of operation there were only 52 infections in all. Wedum reported that both P3 and P4 conditions can reduce infections to extremely low levels, but vaccination is needed to attain virtually complete safety. A research unit working with tularemia, for example, suffered 3.5 infections a year while working on open bench tops, 2.3 infections a year after P3-type biological safety cabinets had been installed, and none at all for the ten years after development of an effective tularemia vaccine.

Wedum concludes that "in the absence of effective vaccination it is not possible to do basic research with a highly infective agent on the open bench top, nor [even with the equipment] specified in P4, without laboratory infections." [8] (There is no vaccine against *E. coli*, but then neither does it seem to be a highly infectious agent.)

Skepticism about relying too heavily on physical containment was often expressed during the formulation of the NIH guidelines. Philip Handler, president of the National Academy of Sciences, remarked on one occasion that he didn't understand what P1 containment really was, or indeed the difference between it and P2: "If you are serious you will go to P3. I don't think P1 and P2 contain anything at all, honestly." [9]

As for P3, Paul Berg cited the fact that even after his laboratory had been upgraded to the P3 level, almost everyone coming to work there acquired substantial anti-body titers to—in other words, had been infected by—the SV40 virus being studied. "Since our experience is the common one, how can anyone be confident that P3 containment will prevent the escape of a potentially dangerous microorganism? Very likely, people working in a P3 containment facility work a bit more carefully but in time familiarity causes a relaxation of effort and mistakes are not uncommon," Berg wrote to the NIH. His conclusion: "I'm convinced that physical containment is over-rated and that P3 containment, while reassuring to the psyche, is hardly the line of defense one would like to put the greatest reliance on. P3 physical containment is vulnerable to human error and, therefore, exposure of personnel within and out of the lab cannot be eliminated; reduced somewhat but not eliminated." [10]

Physical containment, in short, is a variable factor, providing good general protection for the careful researcher, some but considerably less protection for the slovenly. How much protection is enough against *E. coli*? There are not yet any clear answers to the question, but the K12 laboratory strain of *E. coli* is believed to be a less robust organism than the ordinary variety, and the presumed sickliness of K12 provides at least a relative measure of protection. Use of *E. coli* K12 and its crippled cousin, *E. coli* chi-1776, constitute the biological containment system, which is the other pillar of the NIH guidelines.

During the formulation of the guidelines three issues came under particular discussion: the suitability of *E. coli* in general as a host for gene splicing, the degree of enfeeblement of *E. coli* K12, and the permanence of disability of chi-1776.

A not inconspicuous disadvantage of *E. coli* as the basis for a biological containment system is that its natural ecological niche is man. The bacterium inhabits the gut of humans and other warm-blooded animals. Some 10 to 15 percent of the human population also carries it in their nose and throat.[11]

E. coli enjoys a wide distribution in nature.[12] It has been recovered from insects and fishes as well as warm-blooded animals. It flourishes in sewage and can live for long periods in fresh water. In warm, moist soil it can survive for years. It exchanges genetic information with some thirty other groups of bacterial species.

E. coli usually leads a blameless life, doing no harm to its human host. Yet it can cause a variety of diseases. It has been implicated in a number of intestinal illnesses, ranging from infantile diarrhea to traveler's diarrhea to a disease with ulcerative lesions that resembles bacillary dysentery. Infections with E. coli "are probably responsible for the vast majority of diarrheal diseases and other enteric disorders among children and adults in the U.S.A.," according to Roy Curtiss.[13] The bacterium is also one of the three main organisms associated with deaths of hospital patients from septicemia.[14] It is the most common cause of urinary tract infection.

In most, perhaps all, of these disease-causing roles E. coli must first pick up new genetic information, often by acquisition of a plasmid. According to NIH committee member Stanley Falkow of the University of Washington, E. coli "is not ordinarily a highly virulent organism," but the gaining of new genetic information "may be sufficient to tip the balance from a strain that is usually a commensal to one which is capable of initiating overt disease."[15] From the standpoint of recombinant DNA research, Falkow adds, the possibility of E. coli being turned into a disease agent by acquiring a plasmid "must be viewed as one of the most cogent arguments for the potential biohazards associated with this research."

Over the last ten to twenty years the amount of E. coli–caused disease seems to have increased, and so too does the frequency with which plasmids are being found in *E. coli*. Methods of detecting plasmids have also improved in recent years, but the increase seems to be real, and in Curtiss's opinion it represents "a major evolutionarily significant change in the genetic potential of *E. coli* and of other microorganisms as well."[16]

Not much is known about the infectivity of *E. coli*. Most people seem to harbor a semistable population of *E. coli* strains in their intestines, one strain being replaced by another every few months or so. The factors which favor the installation of one particular strain among the many strains that are being continuously ingested are not yet understood. In experiments in which strains of *E. coli* are fed to people it is hard to get any of the new strains to implant. Transmission of *E. coli* occurs by what bacteriologists call a fecal-oral route; carried by the hands or via the preparation of food, the bacteria are ingested and in this way spread from person to person. There is some evidence to suggest that worldwide dissemination of a single serotype, or clone, of *E. coli* can occur rapidly under the right selective pressures.

So much for the free-living strains of *E. coli*; how is the laboratory strain, K12, different? Just as some pathogenic strains of bacteria tend to lose their virulence after being cultured for several generations in the laboratory, so *E. coli* K12 seems to have become adapted to life in the test tube, which began when it was first isolated from a patient in 1922. K12 is customarily said to be enfeebled, although the NIH guidelines do not state this to be the case. According to an English study, the Ashby Report, *E. coli* K12 "is commonly supposed to be a chronically enfeebled strain, but we do not believe that enough is known about its behavior to justify complacency in its use. . . . It would not be prudent, in our view, to assume that the commonly used K12 strain of *E. coli* is enfeebled and quite harmless."[17]

K12 is assumed to lack robustness because in appearance it is what bacteriologists call a "semi–rough" strain. Rough strains, as opposed to smooth, generally do not cause disease; there seems to be something wrong with their cell coat that makes them easy victims for the body's defense mechanism. Unlike some other *E. coli* strains, *E. coli* K12 has never been implicated as the cause of an illness.

An interesting test of this point has been carried out since the Ashby Report by Falkow. He inserted into K12 a pair of plasmids known to make ordinary strains virulent and fed the bacteria to calves. Even the super-K12 failed to make the calves sick.[18]

As far as gene splicing is concerned, the chief risks are, first, that *E. coli* K12 might be turned into a noxious organism by the inserted genes and, second, that it might either colonize a person's gut or transfer the inserted plasmid to a resident strain of *E. coli*, converting it to a pathogen. On present evidence it seems difficult even deliberately to convert K12 into a conventionally pathogenic *E. coli*. As to the chances of K12 colonizing the gut, the tests reported at the Asilomar conference by Ephraim Anderson and H. Williams Smith indicated that the bacterium when fed to volunteers survives for a few days in the gut but does not establish permanent residence. But some microbiologists find these feeding tests unconvincing because it is hard to get any strain of *E. coli*, not just K12, to implant in this way. Rolf Freter of the University of Michigan says of the Anderson and Williams Smith experiments that "to the casual observer this kind of evidence may appear to be quite pertinent and conclusive. Unfortunately this is not the case. . . ." What the evidence does show, according to Freter, is that "K12 has *no greater* ability to establish itself in the human intestine than [do] indigenous strains."[19] Even less is known about K12's ability to survive in other bodily niches, such as the nose and skin. Curtiss has reported that during a period when he took nasal swabs from himself he "rather routinely was able to re-isolate strains of *E. coli* K12 that I was then working with in the laboratory."[20] Data from a single individual do not support any general conclusion one way or the other about K12's behavior in this respect.

Thus while it looks as if K12 finds it difficult or nearly impossible to colonize the gut of a normal healthy person, this supposition has not yet been proved. But even if K12 is unlikely to establish itself, could it transfer its plasmid while en route through the gut to one of the resident strains? The NIH guidelines specify that all plasmids used with K12 must be of a type that is not naturally transmissible from one bacterium to another. Plasmids such as pSC101, one of the standard cloning vehicles, are of this type. It so happens that even nonselftransmissible plasmids can transfer to another bacterium if their host is first infected by a "mobilizing" plasmid. Experimental tests indicate that the chance of a pSC101 plasmid being mobilized in this way are not infinitesimal but are certainly extremely small.[21]

Even if K12 is more sickly than natural strains of *E. coli*, it is far from being completely enfeebled, as NIH committee member Roy Curtiss discovered in trying to make a genetic cripple of it. "As we soon learned," wrote Curtiss, ". . . *E. coli* [K12] had a will as well as a means to survive under adverse conditions when it should have dropped dead. It has required a total of 13 genetic manipulations to obtain a strain that should meet the requirements for an EK2 host as specified in the . . . NIH guidelines."[22]

The strain eventually developed by Curtiss and his team, known as *E. coli* chi-1776, is incapable of synthesizing either its cell wall or the wall undercoat without being fed a diet of special chemicals. Chi-1776 is designed so that should it escape from the laboratory, it would be unable to propagate for lack of the needed chemicals. Some possible snags in this design have been emphasized in a critique by virologist Richard Goldstein and colleagues at the Harvard Medical School; in rather hostile environments—such as tap water for example—chi-1776 survives as well as does *E. coli* K12.[23] Curtiss has pointed out another risk: chi-1776 takes so long to divide—an hour, compared with the twenty minutes for ordinary *E. coli* strains—that it would be rapidly outgrown if the culture were contaminated by a robust *E. coli*. The researcher might proceed with gene insertion, not noticing that his flask of chi-1776 had become a flask of viable bacteria. Such a contamination accident could yield bacteria "with excellent chances for survival and perpetuation of cloned DNA," Curtiss warns.[24]

Whatever the enfeeblement of K12 and the disability of chi-1776, there was a strong difference of opinion during the framing of the NIH guidelines as to whether *E. coli* in any form was a suitable host for gene-splicing research. "It's a tragedy that *E. coli* happens to be picked as the standard [laboratory] organism," remarked NIH committee member Wallace Rowe in 1974, long before the committee had even been thought of. By October 1975, just before the committee agreed on the final draft of its guidelines, Rowe had heard little to make him change his mind. "I still feel strongly that *E. coli* is the wrong host for use with the recombinant DNA molecules. However, I don't think that the people involved in these studies would ever accept a complete moratorium pending development of an alternative host," he wrote.[25]

Another committee member who had doubts about *E. coli* but also saw no opportunity to assuage them was Waclaw Szybalski of the University of Wisconsin. "I don't think *E. coli* is the final answer. I would like to see the use of bacteria and phages that can only grow in hot sulphur springs [and would die on escape from an environment of this high temperature]. The problem is that it takes a long time to develop such things and people don't want to wait," Szybalski remarked in November 1975. Even DeWitt Stetten, the chairman of the committee and NIH deputy director for science, made the same point in a letter to Paul Primakoff of the Harvard Medical School. Stressing that he was not an expert in recombinant DNA research, Stetten wrote:

> You are . . . undoubtedly correct that *E. coli* is the wrong microorganism. Even at the Asilomar conference, however, I detected little interest on the part of the majority to table *E. coli* and begin from scratch with some other organism. The enormous quantity of accumulated information about *E. coli* appeared to dictate that, despite its hazards, this was still the organism of first choice. . . . I should expect that were we to make regulations banning activity in this or any other field of science for a number of years, we should find these regulations very difficult or impossible to enforce.[26]

The *E. coli* system may well constitute a viable method of biological containment, yet in the view of at least some members of the NIH committee, the prime reason for choosing it was not its safety but the judgment that scientists would not accept regulations that prohibited use of the bacterium.*

---

*Would a temporary ban on *E. coli* have been flouted? Would other countries have followed the American lead? The European Molecular Biology Organization at one stage indicated to the NIH it would not go along with the NIH guidelines if the draft version became much stricter,[27] and it would almost certainly have been strongly tempted to ignore a prohibition on *E. coli*. On the other hand, the influence of moral example in this field has consistently proved much weightier than expected. The Academy group's call for a moratorium on certain experiments was heeded worldwide for seven months until the Asilomar conference, and the principles drawn up there were more restrictive than the moratorium yet were accepted unanimously by the foreign delegates (two American scientists cast the only votes against the final statement), which suggests that a majority might have accepted even more stringent conditions. On April 7, 1977, President Carter renounced plutonium-based nuclear power, a technology in which other countries have sunk relatively enormous investments, in the explicit hope of setting an example that the rest of the world would follow. Against this background almost any set of safety conditions the NIH endorsed, however restrictive, would probably have had a fair chance of being accepted worldwide.

Given the doubts within the committee, it is not surprising that a similar debate over *E. coli* took place outside it. The Boston Area Recombinant DNA Group, a coterie of young scientists at the Harvard Medical School and Brandeis University, recommended that the search for an ecologically more appropriate host should start at once and that *E. coli* should be phased out within two years.[28] The social benefits to be gained from gene splicing, the group declared, "will be of equal value whether they come in 20 years versus 25, 50 versus 55, 100 years versus 105."[29]

Defending the NIH position, Paul Berg of Stanford responded to this argument by saying that the search for an alternative host to *E. coli* "could take us 10, 20 or more years, and would we be more secure?" As for delaying the social benefits, "It is pure unadulterated sophistry to argue that a 5 or 10 year delay in achieving the promised advances is acceptable in the long term."[30]

Some critics, however, were not prepared to allow even a two-year phase-out period for *E. coli*. Erwin Chargaff of Columbia, whose research on the chemistry of DNA had laid the groundwork for Watson and Crick's discovery, denounced the choice of *E. coli* as "the cardinal folly." "If our time feels called upon to create new forms of living cells—forms that the world has presumably not seen since its onset—why choose a microbe that has cohabited, more or less happily, with us for a very long time indeed? The answer is that we know so much more about *E. coli* than about anything else, including ourselves. But is this a valid answer? Take your time, study diligently, and you will eventually learn a great deal about organisms that cannot live in men or animals. There is no hurry, there is no hurry whatsoever," Chargaff opined in a magisterial letter to *Science*.[31]

Considerable weight should be given to the views of Curtiss, the biologist who took the concept of biological containment sufficiently seriously that he devoted the full resources of his laboratory for a year to the development of chi-1776. Curtiss was one of the chief enthusiasts of the idea at Asilomar, grew somewhat skeptical during his long struggle to disarm *E. coli* K12, but now believes that the goal has been achieved within reason. Despite *E. coli's* communicability and propensity to virulence, "to choose another microbial host [for gene-spliced molecules] about which less is known, under the expectation that such similar problems do not exist, seems irresponsible to me," Curtiss has written.[32]

Escherichia coli is a natural inhabitant of man, distributed ubiquitously in nature, capable of causing a spectrum of disease from the mildly aggravating to the fatal, and at present undergoing an evolutionary shift in its pathogenicity of unknown significance. Considered from the standpoint of health and ecology alone, it is the worst possible choice as the host for gene-splicing research. If, by some remote chance, a serious accident should eventually result, historians may be moved to inquire why of all the microorganisms at our disposal we had to use *E. coli*. The answer would have to be along the following lines: Prohibition of *E. coli*, by far the best studied of all bacteria, would have delayed gene-splicing research for the several years it would have taken to test out an alternative host. Researchers were unwilling to accept a major delay for what seemed to most to be a purely speculative risk, particularly since the risk was significantly diminished by use of the K12 strain and chi-1776. Research and the pursuit of knowledge being fundamental values of industrialized societies, the majority of the public was favorably disposed to accept the scientists' arguments.

NOTES

1. *Guidelines for Research Involving Recombinant DNA Molecules.* National Institutes of Health, June 1976.

2. Colin Norman, "Maintaining Momentum," *Nature* 266 (17 March 1977): 210.

3. "Microbiology: Hazardous Profession Faces New Uncertainties," *Science* 182 (9 November 1973): 566.

4. Reported by Allan Silverstone of MIT. U.S. Department of Health, Education, and Welfare, *Recombinant DNA Research*, No. (NIH) 76-1138, 1 (National Institutes of Health, August 1976): 285.

5. W. Emmett Barkley, director of Office of Safety Research, National Cancer Institute.

6. R. E. Kissling, "Laboratory-Acquired Infections." In *Biohazards in Biological Research,* p. 70, Cold Spring Harbor Laboratory, 1973.

7. A. G. Wedum, "The Detrick Experience as a Guide to the Probable Efficacy of P4 Microbiological Containment Facilities for Studies on Microbial Recombinant DNA Molecules" (Paper prepared under contract to the National Cancer Institute, 20 January 1976).

8. Ibid.

9. U. S. Department of Health, Education, and Welfare, *Recombinant DNA Research*, p. 325.

10. Berg to DeWitt Stetten, 2 September 1975, in files of NIH Recombinant DNA Molecule Program Advisory Committee.

11. Working Party on Potential Biohazards Associated with Bacterial Plasmids and Phages, Report to the National Academy of Sciences Committee on Recombinant DNA Molecules, 24 February 1975, p. 24.

12. Stanley Falkow, "The Ecology of *Escherichia Coli*" (Memorandum to NIH Recombinant DNA Molecule Program Advisory Committee, included in minutes of meeting on 12–13 May 1975).

13. Roy Curtiss, Memorandum to the National Academy of Sciences Committee on Recombinant DNA Molecules, 6 August 1974, p. 6.

14. Falkow, "Ecology of *Escherichia Coli.*"

15. Ibid.

16. Curtiss, Memorandum to NAS Committee on Recombinant DNA Molecules.

17. *Report of the Working Party on the Experimental Manipulation of the Genetic Composition of Microorganisms (Ashby Report)* (London: Her Majesty's Stationery Office, 1975), pp. 8 and 15.

18. Falkow, "Ecology of *Escherichia Coli.*"

19. Freter to John R. Seal, 7 August 1975, in files of NIH Recombinant DNA Molecule Program Advisory Committee.

20. Curtiss, Memorandum to NAS Committee on Recombinant DNA Molecules.

21. Falkow, "Ecology of *Escherichia Coli.*" Falkow calculates that, using a nonselftransmissible plasmid, such as pSC101, as the cloning

vehicle, the chances of an ingested *E. coli* K12 strain transferring its recombinant DNA molecule to another bacterium would probably not exceed one chance in a million million per gram of feces in the gut per day.

22. Roy Curtiss, "Genetic Manipulation of Microorganisms: Potential Benefits and Biohazards. *Annual Review of Microbiology* 1976.

23. Richard Goldstein, Cristian Orrego, and Philip Yuderian, Analysis and Critique of the Curtiss Report on the *Escherichia Coli* Strain Intended for Biological Containment in DNA Implantation Research.

24. Curtiss, "Genetic Manipulation of Microorganisms."

25. Rowe to Paul Primakoff, 6 October 1975, in files of NIH Recombinant DNA Molecule Program Advisory Committee.

26. Stetten to Primakoff, 6 October 1975, in files of NIH Recombinant DNA Molecule Program Advisory Committee.

27. Charles Weissman, chairman of EMBO Standing Advisory Committee on Recombinant DNA, to DeWitt Stetten, 18 February 1976, in files of NIH Recombinant DNA Molecule Committee.

28. Boston Area Recombinant DNA Group, Memorandum to DeWitt Stetten, 24 November 1975.

29. Ibid.

30. Berg to Donald S. Fredrickson, 17 February 1976, in files of NIH Recombinant DNA Molecule Program Advisory Committee.

31. "On the Dangers of Genetic Meddling," *Science* 192 (4 June 1976): 938.

32. Curtiss, "Genetic Manipulation of Microorganisms."

8
THE FROG AND FRUITFLY FRACAS

THE SYSTEM OF physical and biological containment devised by the NIH committee deserves to be judged on its merits. But among the factors to be weighed in that judgment is the process by which the NIH guidelines were framed. No political procedure is perfect, but the NIH committee on recombinant DNA research came under an unusual barrage of criticisms, many of which continued to reverberate long after it had finished its work.

The intensity of public concern about gene splicing had not become apparent, and evidently was not anticipated, at the time the committee was set up. Its role, like that of many other NIH committees, was to enable a particular group of specialists to solve their own problem in their own way. The approach would have been acceptable and appropriate if no other groups had considered their interests affected. When that proved not to be so, the NIH committee became vulnerable to charges of conflict of interest and narrowness of representation.

Of the twelve members of the original committee, all but the chairman had a judgmental conflict of interest in that as biological researchers they might one day wish to undertake the experiments they were regulating. Three members also had a personal conflict in that they were already engaged in gene-splicing research.[1] Two of these three, David Hogness of Stanford and Charles Thomas of the Harvard Medical School, were the dominant characters on the committee, and both consistently argued in favor of less stringent conditions of containment.

It was in fact to Hogness that the committee entrusted the task of preparing the first draft of the guidelines. The arrangement prompted one critic, Jonathan King of MIT, to declare that the committee's function "is to protect geneticists, not the public" and to compare Hogness's position with "having the chairman of General Motors write the specifications for safety belts." Hogness rejected the charge of conflict of interest, saying that he knew of no disagreement on the safety precautions appropriate for his own area of research—shotgun experiments with the *Drosophila* fruitfly.

Despite the potential for conflict, there has been no suggestion that any committee member acted from conscious bias or argued from anything other than his best judgment. Even King, one of the committee's harshest critics, accused it only of acting on its beliefs that the risks were small, as he made clear in the hearing held before Mayor Vellucci of Cambridge, Massachusetts, the day the NIH guidelines were issued:

> **King:** That was a group of people who were essentially the protagonists; they were the ones doing the experiments. . . . Those guidelines are like having the tobacco industry write guidelines for tobacco safety.
>
> **Vellucci:** Dr. King, are you suggesting this was a stacked deck?
>
> **King:** It was not a stacked deck.
>
> **Vellucci:** It was?
>
> **King:** No, it was not. In hindsight it was a stacked deck. It wasn't intentional. There wasn't any conspiracy. They mean well, they're working hard,

they're trying to write good guidelines. However, since they believe the stuff isn't very dangerous, it's not surprising that the guidelines they write are a little bit lax compared to what many others believe.[2]

King's objections could have been satisfied by constituting the committee with a broader membership, a point that he and other members of Science for the People, an activist political group, had urged even at the Asilomar conference. "Decisions at this cross-roads of biological research must not be made without public participation. . . . Yet we see even in the structure of this conference that a scientific elite is here alone trying to determine the direction that such regulation should take," declared an open letter from the Genetic Engineering Group of Science for the People.

Members of the Genetic Engineering Group were not the only people concerned about who should make the decisions. The NIH committee itself at its first meeting, the day after the Asilomar conference, decided that the range of scientific disciplines represented on it was too narrow and suggested that it should be joined by an epidemiologist, animal virologist, and plant pathologist. The committee also "specifically recommended" that one lay member be appointed.[3]

The recommendation for a member of the public was repeated at its next meeting, in May 1975. Two new members joined the committee shortly thereafter, but both were scientists. The committee continued to press for a nonscientist member, foreseeing its vulnerability to criticism on this score. "Like many other present members of the committee," Jane Setlow of the Brookhaven National Laboratory wrote to the NIH in September 1975, "I'm not sure this person could contribute to the deliberations, but I *am* sure that we need one for the purpose of being able to say we have one when there are complaints."[4]

The committee's desire for a token public member

was eventually addressed by the appointment of a distinguished professor of public administration Emmette S. Redford, of the University of Texas. He attended the meeting of December 1975, at which the committee completed its final draft, but contributed nothing of substance.

The committee's final draft was discussed at a public hearing convened in February 1976 by NIH director Donald S. Fredrickson. Scientific critics, as well as consumers and environmentalists, were invited and raised several major criticisms, such as questioning whether *E. coli* was an appropriate host and whether gene splicing risked upsetting the evolutionary balance. Fredrickson passed on suggestions for only minor changes to the committee, which rejected most of them. The final version of the guidelines, issued in June 1976, thus could not be said to embody any substantial degree of public input. The committee did take advice from specialist consultants, such as plant pathologists, but it never got the epidemiologist member it had requested. This contrasted with the approach taken in England, for example, where the view was that since hazard to public health was the issue, it should be resolved by a committee of medical microbiologists and public health experts with no personal interest in using the technique. The principal defect of the English approach was that its committee met behind closed doors, so that the public had no opportunity of learning the rationale behind the measures taken for its protection. By contrast, all meetings of the NIH committee, as well as its minutes and correspondence, were open to press and public.

Narrowness of composition was perhaps a flaw in the NIH committee, but the test of its labors is how well it accomplished its goals. The committee's formal purpose was in fact "to recommend programs of research to assess the possibilities of spread of specific DNA recombinants and the possible hazards to public health and the environment; and to recommend guidelines on the basis of the research results."[5] In other words, the committee was first, to measure the risk and second, to contain it. Other groups, such as a committee of the European Molecular Biology Organization, have drawn up programs of research to assess the risk, but the NIH committee has not done so. From its first meeting it decided to recommend guidelines and worry about their basis later. The decision was understandable: it was better that the NIH committee should translate the general principles of Asilomar into a code of practice rather than having each laboratory make its own interpretation. The committee did not attempt, however, to perform its two mandated tasks in parallel. The NIH initiated some experiments to assess risk in March 1977.

How religiously did the committee translate the Asilomar principles into practice? There were two slightly different versions of the principles. The original document agreed to by the 140 international delegates stated that the research should proceed under appropriate safeguards but that "it would be wise to exercise the utmost caution." The official account published by the conference organizing committee substituted for "utmost caution" the less arduous exhortation to "considerable caution."[6]

The change was understandable: "Utmost caution," literally applied, would have reduced progress to a snail's pace, which was probably not the intention of the conference. A more substantive change concerned health monitoring of researchers. The English Ashby Report, completed before the Asilomar conference, placed heavy emphasis on monitoring as a way to test if the proposed safety measures were working in practice. "We recommend that such monitoring should become a routine practice and that it should be begun without delay on an international basis," the English report stated.[7] Echoing this position, the delegates at Asilomar agreed to the following statement: "It is strongly recommended that appropriate health surveillance of all personnel, including serological monitoring, be conducted periodically to establish a base for epidemiological analyses."

The published version of the Asilomar statement omitted both the *strongly* and the important phrase *to establish a base for epidemiological analyses.* Asked by what authority the organizing committee made these changes, chairman Paul Berg of Stanford has said that no relaxation of standards was intended and that he sees "no substantial difference" in the changes. But possibly because of those attenuations, the NIH committee did not take steps to establish a base for epidemiological analyses or even to have the health of researchers periodically surveyed. The guidelines leave it to the scientist in charge of each experiment to decide what monitoring needs to be done.

Another aspect in which the NIH committee may have loosened the Asilomar principles a little is in the matter of biological containment. Delegates at the conference were perhaps overconfident that disabled bacteria and plasmids could be developed in a matter of weeks, only Cohen warning that the task would be like "waiting for the Messiah." Thus the Asilomar statement recognizes only one level of biological containment and requires it to be used in all P3 and P4 experiments, and in P2 as soon as the disabled organisms become available. To which level in the NIH guidelines does the Asilomar level of biological containment correspond? Roy Curtiss, while in the throes of trying to disable *E. coli* K12, argued that it corresponded to the EK3 of the guidelines, the fully tested disarmed host. "I would like to eliminate the EK2 category," he wrote, "in that throughout the Asilomar document it is specified that investigators relying upon disarmed hosts and vectors for additional safety must rigorously test the effectiveness of these agents before accepting their validity as biological barriers."[8] Under this severe interpretation (from which Curtiss later receded) the Asilomar document would have allowed use of *E. coli* K12 in experiments suitable for P1 and P2 physical containment, but only EK3 hosts, which are not yet available, would be allowed for P3 and P4 work.

There was at any rate room for disagreement on the interpretation of the Asilomar principles, as the NIH committee discovered on the release of its first draft guidelines in July 1975. Protests flooded in from other scientists as well as committee members who had not attended the meeting. Fifty biologists attending a meeting at Cold Spring Harbor in August signed a petition to the NIH urging that the guidelines be made more stringent and the committee's membership be broadened to include other disciplines and more scientists without an involvement in gene-splicing experiments. "The present draft appears to lower substantially the safety standards set and accepted by the scientific community at Asilomar," the petition declared.[9]

Asilomar organizing committee chairman Paul Berg thought much the same. The levels of containment required for some experiments were "marginal and inadequate," while other features of the draft "are very likely to draw the charge of self-serving tokenism," Berg wrote to the NIH.[10]

Committee member Curtiss was even more critical. He described the draft guidelines as "a license to experiment . . . in the almost complete absence of controls and/or sanctions."[11] Nor did Stanley Falkow of the University of Washington, the committee's only public health expert, hold the draft in particularly high esteem. Some aspects he judged to be "built on a foundation of quicksand," others were "lip service plain and simple" or showed "a certain lack of appreciation of bacterial pathogenesis." As to the containment levels recommended for shotgun experiments with frogs and fruitflies (P2 + EK1 in the draft; frogs are now at P2 + EK2), Falkow described them as "tantamount to a hunting license for any hack or high school student to do these experiments with the blessing of the NIH." [12]

Faced with the barrage of criticism from within and without, the NIH committee met again in December 1975, in the Calypso room of the La Valencia Hotel in La Jolla, California. Whether by design or accident, three members of the Asilomar organizing committee—Berg, Maxine Singer of NIH, and Sydney Brenner of Cambridge, England—attended the meeting as observers.

After some of the minor laxities of the first draft had been corrected, the chief point at issue was the grading of shotgun experiments. The subject was of personal interest to several members of the committee and their scientific colleagues, who stood to be put at least temporarily out of business if the containment levels were raised above those of the facilities they had available.

There are two principal reasons for apprehension about shotgun experiments, one specific and one general. The specific reason is that the gene set of the organism being shotgunned might contain a repressed tumor virus that, on liberation, could be harmful to man. The conjecture is that the closer the species is to man on the evolutionary tree, the more likely are its viruses to present a threat. On this conjecture is based the rationale for grading eucaryotic shotgun experiments into different levels of containment, the human shotgun being assigned to the highest level of containment, the fruitfly shotgun to the lowest.

The general reason for concern about shotguns is simply that they venture through a very large amount of unknown genetic material, the properties of which when inserted into clones cannot be definitely predicted in advance. The difference between this rationale and the tumor virus conjecture is more than merely academic; the practical consequence of the general rationale is that all shotgun experiments are equally hazardous or equally free from hazard and should be assigned to the same level of containment.

The point was most clearly articulated by Brenner: "I think that one has to consider the hazards arising from the insertion of random DNA. What one must avoid is the creation of any selective advantage in any element of a microbe which in the world outside would be of adverse medical or economic significance. We don't know what the probability of that is. The essence of a shotgun experiment is that it explores a very large sample. That is the same whether you use *Bacillus subtilis*, *Drosophila* [fruitflies], or humans. So the production of the hazard is uniform, and should have one level of containment."

The same argument was advanced by a group within the committee favoring stricter guidelines. "If recombinant DNA molecules are in fact hazardous, it is by no means clear that the dangers will be eliminated by using DNA from lower eucaryotes," the members opined. Their solution: put all shotguns in P3 + EK2, except for primates, which should be higher.

But the course of debate at the La Jolla meeting followed the tumor virus rationale for shotguns. "I think we are putting these things extremely high. I think the chances are similar to being struck by lightning," Hogness remarked of the putative dangers of shotgun experiments. The committee voted by 9 to 4 to keep the frog shotgun, a benchmark standard of the debate, at the level of P2 + EK1. The decision lead Brenner to remark privately that to make the containment levels any lower, "you will now have to go to P1 and do the experiment in your boots." Berg observed that P2 + EK1 "is working in your garage."

On the second day of its meeting, by some curious

alchemy, the committee changed its collective mind. John Littlefield of Johns Hopkins Hospital reopened the issue of frog shotguns by saying he would like to raise them to P2 + EK2. The motion was passed by 7 votes to 6, whereupon Hogness at once suggested that if the cells were taken from frog embryonic tissue (supposedly freer of viruses than adult cells), the containment level could be reduced back to P2 + EK1. The motion passed. Later in the day Hogness again raised the issue, suggesting that shotguns with frog adult tissue as well be allowed in P2 + EK1, but lost by the same 7-to-6 vote.

The battle of the frogs demonstrated how particular interests as well as general principles shaped the framing of the NIH guidelines. Another demonstration was that shotgun experiments with higher plants, which on the tumor virus rationale are rather considerably less threatening to man than are fruitflies, were left in the stricter containment level of P2 + EK2. Why? Because no one at the committee was particularly anxious to shotgun higher plants, whereas many researchers were interested in frogs and fruitflies. Higher plants have since been downgraded to the same level as fruitflies.

Brenner advised the committee that it should spell out clearly in its report the rationale for assigning each kind of shotgun experiment to a particular level of containment. "To people from outside this thing looks like the settling of all sorts of different bargains. That may sound obnoxious, but that is how it looks," Brenner said.

Brenner's advice was not taken, and the NIH guidelines in consequence are a Byzantine document that gives the reader only hints as to either the political or scientific rationales for the maze of rules it describes.

Yet neither rationale is discreditable. In political terms, the guidelines agreed upon at La Jolla were a compromise between opposing views but one that was accepted unanimously. Those who favored more stringent guidelines, such as Wallace Rowe, Curtiss, and Falkow, believed the compromise was acceptable, and Berg, who had been the harshest critic of the earlier draft, pronounced the revision to be "a faithful translation of the spirit of Asilomar." The compromise was also tolerable to those who believed the guidelines were stricter than necessary, such as Hogness, Charles Thomas, and James Darnell. Darnell, a molecular biologist at Rockefeller University, remarked afterward that the guidelines, though more stringent than they need be, "are no serious impediment to research."

The scientific rationale of the NIH guidelines rests in large measure on guesswork and judgment about unknown risks. There is general agreement that the hazards that the guidelines are designed to preclude are not likely to occur. The physical and biological systems of containment should provide a substantial measure of safety, even if in the hands of less skilled researchers who work considerably less perfectly than intended.

If based on miscalculation, the guidelines can always be remedied at the first sign of inadequacy. Only if some sudden catastrophe occurs, or some imperceptible danger that is not noticed until too late, will the consequences of error be irreparable.

Yet even if the NIH guidelines provide a reasonable bulwark against the perceived hazards, with a margin of safety to spare, what of the hazards the committee did not take into account? After the guidelines had been finalized at the La Jolla meeting in December 1975, another set of objections was raised by a new wave of critics, foremost amongst whom were Erwin Chargaff and Robert Sinsheimer.

NOTES

1. Jane K. Setlow to DeWitt Stetten, 4 September 1975, in files of NIH Recombinant DNA Molecule Program Advisory Committee.

2. Hearing on Recombinant DNA Experimentation, 23 June 1976 (Cambridge City Council transcript), p. 135.

3. Minutes of meeting, 28 February 1975, NIH Recombinant DNA Molecule Program Advisory Committee.

4. Setlow to Stetten, 4 September 1975.

5. *Federal Register* 39 (6 November 1974): 39306.

6. "Asilomar Conference on Recombinant DNA Molecules: Summary Statement of Report Submitted to the Assembly of Life Sciences of the National Academy of Sciences," *Science* 188 (6 June 1975): 991.

7. Report of the Working Party on the Experimental Manipulation of the Genetic Composition of Microorganisms (Ashby Report) (London: Her Majesty's Stationery Office, 1975), p. 8.

8. Curtiss to DeWitt Stetten, 22 September 1975, in files of NIH Recombinant DNA Molecule Program Advisory Committee.

9. Letter to DeWitt Stetten, 27 August 1975, in files of NIH Recombinant DNA Molecule Program Advisory Committee.

10. Berg to DeWitt Stetten, 2 September 1975, in files of NIH Recombinant DNA Molecule Program Advisory Committee.

11. Curtiss to DeWitt Stetten, 13 August 1975, in files of NIH Recombinant DNA Molecule Program Advisory Committee.

12. Comments, undated, in files of NIH Recombinant DNA Molecule Program Advisory Committee.

9

HEROSTRATUS
AND THE FLAG
OF GALILEO

SCIENTISTS TOOK the initiative in drawing attention to the possible hazards of gene splicing and continued to provide the severest critics of the research from within their own ranks. But this internal debate, admirably frank and freewheeling, caused increasing strains within the scientific community as the consensus began to show cracks in the course of 1976 and as the public, attracted by the noise of the discussion, took a belated interest in what was going on.

The public debate, by and large, has been a replay of the issues raised in the debate among scientists, in which at least four different stages can be distinguished. Those behind the Dieter Söll–Maxine Singer letter of 1973, and the Academy group that invoked the moratorium, opened the first stage of the debate. The second stage was marked by scientists' reaction against the NIH committee's first draft of guidelines. Among the assailants were Singer and Paul Berg from the first wave of critics, and the Boston Area Recombinant DNA Group, which acted in part at the urging of several members of the NIH committee who felt certain issues were being railroaded.

The unanimous decision reached at the committee's

meeting in December 1975 seemed to restore consensus, but then came the third wave of critics, constituted by Erwin Chargaff and Robert Sinsheimer, two eminent scientists who found fault not with the details but with the whole approach of the NIH guidelines. Their intervention paved the way for the fourth wave of critics, a group of politically committed scientists at Harvard and MIT. It was the activities of this group, expressed in the spring and summer of 1976 through discussions within Harvard and at the public forum provided by Mayor Alfred Vellucci of Cambridge, that began the politicization of the debate and drew public attention to the discords among scientists.

The spread of the debate to public arenas was regarded with a mixture of vexation and foreboding within the scientific community. Dependent on public patronage but distrustful of their patrons' true devotion to pure research, scientists feared that the public would fail to understand the subtleties of the arguments involved, that it would be unmoved by the excitement of the avenues opened up by the technique, and that, faced with even an irreducible minimum of risk, it might severely restrict or shut down the research altogether.

Under the pressure of such apprehensions, even the Academy group found itself under attack from less civically minded scientists for having made a public issue of gene splicing. "Some of us are now in the awkward position of having our colleagues ask us, 'What have you done to the future of genetic research?' " complained Norton Zinder of Rockefeller University.[1] "There are people who say, 'If you guys hadn't opened your mouth, nothing would have happened, it would all have blown away.' "

James Watson, another member of the Academy group, viewed the growing debate over gene splicing with undisguised horror. "I told Sargent Shriver that recombinant DNA is the most overblown thing since his brother-in-law created the fall-out shelter debacle," Watson announced to a public hearing held in October 1976 by the attorney general of New York State. "What started out as an attempt by the scientific community to appear responsible takes on increasingly the aspects of a black comedy. I was President Kennedy's adviser on biological warfare. I knew all we had at Fort Detrick, and if I can reveal a secret about what we had, what we had was nothing. The marginal danger of this thing is a joke compared to [even what we had at Fort Detrick]."

As the black comedy unfolded during 1976 and early 1977, with more public hearings, TV programs, and magazine articles treading round the same path, there was a noticeable hardening of opinion within the scientific community. Battle lines formed around the NIH guidelines, although there remained a wide spectrum of opinion among both the proponents and opponents of the approach that the guidelines embodied. The language on both sides grew noticeably shriller. Chargaff started it by referring to his colleagues as "feeble men, masquerading as experts."[2] Berg accused the Cambridge critics of "sophistry bordering on dishonesty."[3]

Despite the volume of the debate, the number of scientists actively involved on either side has remained rather small. The task of defending the approach of the NIH guidelines in public arenas has been assumed mostly by Maxine Singer of the NIH, by Berg and other members of the Academy group, by NIH director Donald Fredrickson, and by members of the NIH committee. The opponents consist chiefly of Chargaff and Sinsheimer, and the Cambridge-based critics.

As for the silent majority of the scientific community, a casual impression would tend to support the proponents' claim that they speak for almost all other scientists. On the other hand, two opinion surveys taken a year apart indicate that a surprisingly large minority of biologists seem to favor stricter guidelines.

A straw poll among its members was taken in April 1976 by the Federation of American Scientists, a politically moderate group whose chief interest is arms control. Of the 129 biologists who replied, 74 percent said the NIH guidelines were probably about right or too restrictive, whereas 25 percent deemed them "probably insufficiently cautious."[4]

A more systematic survey was conducted in February and March 1977 by Robert Cooke, science correspondent for the *Boston Globe.* Cooke mailed questionnaires to a random selection of some 1,250 senior experimental biologists working in universities, industry and government and received 490 replies. "A large number of scientists— 39 percent—believe the rules [the NIH guidelines] should be made even more strict, while 44 percent think the new guidelines are strong enough," Cooke reported.[5]

The 39 percent reported as favoring stricter guidelines is a remarkably large minority. The number may in fact be overstated because the question apparently answered by the respondents asked if they wanted "research to continue under stricter rules, extended to cover industry." Some respondents may merely have wanted the already existing rules extended to cover industry. Yet assuming a spectrum of opinion, many must also have supported stricter rules as well because no less than 10 percent of the 369 academic scientists said they wanted a temporary moratorium on all research and 1 percent (3 individuals) opted for a complete ban.

Eighty-two of the biologists described themselves as "very well informed" about the issue of recombinant DNA research. Among this group, Cook reports, the greatest number said work should continue under the existing NIH guidelines, a "substantial minority" favored stricter rules extended to cover industry, 5 percent recommended a temporary moratorium, and none demanded a complete ban.

The third wave of critics, Chargaff and Sinsheimer, may well have been responsible for some of the doubts that still linger about the approach of the NIH guidelines. Science is an elitist institution, and in certain matters scientists heed who is saying it no less than what is said. Had the call for a moratorium come from scientists less well known than those in the Academy group, it would probably have fallen on very stony ground. Similarly, when Chargaff and Sinsheimer raised objections of principle to the NIH approach, they were probably accorded a careful hearing. Both are members of the National Academy of Sciences and have made discoveries that merit mention in the basic textbooks of molecular biology.

Chargaff laid the chemical groundwork, just as Rosalind Franklin established the crystallographic basis, for the discovery of the structure of DNA. The Nobel Prize for the discovery went to Watson and Crick for their acuity in fitting those two and other pieces of data together. A few of Chargaff's colleagues are wont to interpret his astringent prose style and mocking essays as an expression of bitterness at not having shared in the Nobel Prize. A more plausible explanation of Chargaff's outlook on the world is that, while watching European civilization collapse from 1914 onward, he became a devotee of the mordant Viennese satirist Karl Kraus. "Even when I was young, my inclinations always were in favor of critique and scepticism," Chargaff has written, extending characteristically soothing regrets to the victims of his tail tweaking: "Nevertheless, if at one time or another I have brushed a few colleagues the wrong way, I must apologize: I had not realized that they were covered with fur."[6]

Chargaff's first entry into the recombinant DNA debate occurred in an essay adumbrating one of his favorite themes, the perversion of modern science from an intellectual into an entrepreneurial pursuit. "Knowing that the desire to improve mankind has led to some of the most horrible atrocities recorded by history," Chargaff wrote,

> It was with a feeling of deep melancholy that I read about the peculiar conference that took place recently in the neighborhood of Palo Alto. At this Council of Asilomar there congregated the molecular bishops and church fathers from all over the world, in order to condemn the heresies of which they themselves had been the first and the principal perpetrators. This was probably the first time in history that the incendiaries formed their own fire brigade.[7]

A few months later, in a letter to *Science* entitled "On the Dangers of Genetic Meddling," Chargaff let slip his satirist's mask, revealing the moralist underneath:

> What seems to have been disregarded completely is that we are dealing here much more with an ethical problem than with one in public health, and that the principal question to be answered is whether we have the right to put an additional fearful load on generations that are not yet born. . . . You can stop splitting the atom; you can stop visiting the moon; you can stop using aerosols; you may even decide not to kill entire populations by the use of a few bombs. But you cannot recall a new form of life. Once you have constructed a viable *E. coli* cell carrying a plasmid DNA into which a piece of eukaryotic DNA has been spliced, it will survive you and your children and your children's children. An irreversible attack on the biosphere is something so unheard of, so unthinkable to previous generations, that I could only wish that mine had not been guilty of it. The hybridization of Prometheus with Herostratus* is bound to give evil results. . . .
> Have we the right to counteract, irreversibly, the evolutionary wisdom of millions of years, in order to satisfy the ambition and curiosity of a few scientists?
> This world is given to us on loan. We come and we go; and after a time we leave earth and air and water to

---

*Prometheus stole the gift of fire from the gods and gave it to man; Herostratus burnt down the temple of Diana at Ephesus to give himself a name in history.

others who come after us. My generation, or perhaps the one preceding mine, has been the first to engage, under the leadership of the exact sciences, in a destructive colonial war against nature. The future will curse us for it.[8]

Most gene-spliced organisms that escape from the laboratory or factory will perish in nature, and almost all will die in the long run—most species that have ever lived are now extinct. But Chargaff is correct in the sense that a laboratory-created organism that happened to find a niche in nature could probably not be eradicated by man; the change would be irreversible.

Chargaff's argument, however, is not about technical issues. He is making a purely moral judgment, based in part on an aesthetic consideration of what science should be about. "We were told, when I was a student," Chargaff remarked at a symposium held in Brooklyn College, "that science was one of the attempts of humanity to learn the truth about nature. We were not told to go ahead and improve on nature or to modify nature. We were going to understand it."

Gene splicing was invented as a tool for understanding nature; it is just an incidental side effect, as far as the basic researcher is concerned, that it happens also to change nature. Most people would say at that point that there is nothing inherently wrong in altering nature—human societies have been doing so in other ways for a long time—and that the technique offers the promise of great progress in knowledge about basic biological mechanisms. Chargaff, however, does not applaud all forms of progress. "I am one of the few people old enough to remember that the extermination camps in Nazi Germany began as an experiment in genetics," he remarks. His case against gene splicing is that of a pessimist who fears that the knowledge may be misused, of a humanist who believes that the march of science should where necessary be constrained by other values, and of a moralist who is persuaded that the laboratory creation of new forms of life is simply wrong.

So conservative an outlook is unlikely to prove generally persuasive, but then Chargaff does not expect it to be. Borrowing Cassandra's mantle, he announced to a forum sponsored by the National Academy of Sciences in March 1977: "Taking a historical view, I shall merely point out that minorities are often vindicated by future events, but never before it is too late."

The other critic of major stature within the scientific universe is Robert Sinsheimer, chairman of the division of biology at Caltech. A member of the National Academy of Sciences and editor of its *Proceedings*, Sinsheimer's career as a biophysicist has included study of the nucleic acids of viruses and in particular of a scientifically interesting virus known as phi-X174.

"To impose any limit upon freedom of inquiry is especially bitter for the scientist whose life is one of inquiry; but science has become too potent. It is no longer enough to wave the flag of Galileo," Sinsheimer remarked in a recent lecture.[9] Yet he has not always been so skeptical of the fruits of scientific progress. As his opponents find frequent occasion to recall, he was once an ardent advocate of genetic engineering. In an article written in 1970 he looked forward to the advent of human genetic engineering as a way to escape the tyranny of heredity and to improve man's intellect and other capacities.[10]

Sinsheimer now believes otherwise. In a recent talk he warned of the dangers that may accompany new knowledge: " 'Know the truth and the truth shall make you free' is a credo carved on the walls and lintels of laboratories and libraries across the land." But,

> We begin to see that the truth is not enough, that the truth is necessary but not sufficient, that scientific inquiry, the revealer of truth, needs to be coupled with wisdom if

our object is to advance the human condition. . . .

The twentieth century has seen a cascade of magnificent scientific discoveries. Two, in particular, have extended our powers far beyond prior human scale and experience. In the nucleus of the atom we have penetrated to the core of matter and energy. In the nucleic acids of the cell we have penetrated to the core of life.

When we are armed with such powers I think there are limits to the extent to which we can continue to rely upon the resilience of nature or of social institutions to protect us from our follies and our finite wisdom. Our thrusts of inquiry should not too far exceed our perception of their consequence. There are time constants and momenta in human affairs. We need to recognize that the great forces we now wield might—just might—drive us too swiftly toward some unseen chasm.[11]

For a scientist, that is some change of faith. What has converted Sinsheimer from advocate to skeptic, from enthusiasm about genetic engineering to misgivings so grave as to set him on a different path from the mass of his colleagues?

In his earlier view of genetic engineering Sinsheimer explains, "I thought of very careful experiments to replace gene A with gene B—it never occurred to me that anyone would do a shotgun experiment [in which all the genes of an organism are manipulated at random]."[12] He was also more optimistic that genetic engineering could be controlled.

Asked why more of his colleagues do not share his apprehensions, Sinsheimer replies, "I have been thinking about these things for longer than have most people who are now more sanguine than I am. Scientists can be very insular, and to some degree they have to be. To be a good scientist takes an awful lot of dedication, and you really have to believe in it and believe that what you are doing is good and beneficial. It is such people who are less likely to entertain other points of view."

A certain narrowness of view is one of the complaints Sinsheimer has about the NIH committee's guidelines governing gene-splicing research. "This is a technology that was developed by scientists to solve their own problems, and they are still locked into that mode of thinking."

On reviewing the guidelines at the NIH's request, Sinsheimer found that they had dealt reasonably well with the immediate health hazards but "had given no thought to the evolutionary question."

The oversight was not in his view surprising: "It was implicit for the guidelines committee to concern itself with health hazards—it simply was not constituted to cope with the larger issues." Part of the fault is that most biological research in the United States is financed by the NIH, an agency whose primary mission is health. This dependence, Sinsheimer considers, has distorted biology and biased scientists' values. For lack of an evolutionary perspective, the NIH guidelines

> reflect a view of Nature as a static and passive domain, wholly subject to our dominion. They regard our ecological niche as wholly secure, deeply insulated from potential onslaught, with no chinks or unguarded section or perimeter. I cannot be so sanguine. In simple truth just one —just one—penetration of our niche could be sufficient to produce a catastrophe.[13]

Living species, as Sinsheimer often points out, are all the product of some three billion years of evolution. The evolutionary tree that traces the development of all species of life represents the fact that evolution proceeds in a linear, branching manner to produce gradually diverging species. The cardinal feature of the tree is that its branches do not cross or interweave. Nature has so arranged matters that, by and large, genes only interact within a particular species.

The gene-splicing technique enables man to turn the tree into a network by merging genes from one species with those of another. The technique also fractures another of nature's arrangements: the measured pace by which, in a series of gradual interactions, a new species enters the quasi-equilibrium of natural ecology or an old species fades out. Man can now create a new species from nowhere and dump it into the system without notice. As Sinsheimer puts it, "Now we come, with our splendid science and our ingenuity, and we have now the power to introduce quantum jumps into the evolutionary process. But do we have the commensurate understanding to foresee the consequences to the currently established equilibria on which, quite literally, our life support systems depend?"

In a shotgun experiment, for example, the DNA from an insect or sea urchin can be cut with a restriction enzyme into some fifty thousand fragments, say, each containing an unknown cluster of genes. Using a second restriction enzyme, with an affinity for a different sequence of bases, another set of fifty thousand fragments can be produced for insertion into bacteria and cloning. Says Sinsheimer: "Somehow it is presumed that we know, *a priori*, that none of these clones will be harmful to man or to our animals or to our crops or to other microbes— on which we unthinkingly rely. I don't know that and, worse, I don't know how anyone else does."[14]

Shotgunning fruitfly or sea urchin genes into *E. coli* is an example of a general procedure that Sinsheimer fears may hold particular hazards: that of inserting the genes of eucaryotes (the cells of higher plants and animals) into procaryotic cells (bacteria and certain algae). Though procaryotes and eucaryotes interact intensely with each other as organisms, they are not known to interact with any frequency on the genetic level.

A possible reason for this lack of intercourse is that procaryotes and eucaryotes, although they use the same genetic code, may differ in the control elements and genetic signals that govern how the code is put into operation. The danger of putting any piece of eucaryotic DNA into a procaryote is that it may endow procaryotes with the eucaryotic control signals, a kind of betrayal of state secrets at the molecular level.

Even if this should occasionally happen in nature, as when a bacterium is infected by DNA from the debris of a eucaryotic cell, numerous experiments of the type now being conducted in laboratories all over the world may significantly increase the risk, Sinsheimer says.

What might be the consequences of breaching the natural barrier that appears to exist between procaryotes and eucaryotes? Perhaps the viruses that now prey on bacteria might acquire the genetic machinery to infect eucaryotic cells. Perhaps bacteria might acquire the genes to serve as reservoirs for some of the viruses that infect man or other animals. "One need not continue to spin out horror stories," Sinsheimer has written: "The point is that we will be perturbing, in a major way, an extremely intricate ecological interaction which we understand only dimly."[15]

The difficulty of assessing Sinsheimer's apprehensions is that evolutionary biologists have not hitherto paid much attention to the nature of the procaryote-eucaryote barrier and have little hard information to bring to bear on the subject. Many biologists disagree vehemently with Sinsheimer's barrier theorem, but the strength of the reaction seems sometimes to relate to fear that his views will be used to impede research. Since raised only in response to his objections, there is an unavoidably *ad hoc* element to some of the arguments brought against his theorem.

One such argument is that the barrier may have no particular purpose, having just arisen accidentally as procaryotes and eucaryotes took their separate paths. Another supposition is that the barrier doesn't exist at all, it just seems to exist because we are ignorant of the flow of genes between species.

A variant of this theme has it that the barrier is being

broken all the time in nature, as for example when bacteria in the gut take up fragments of DNA from digested cells. Why then have altered bacteria not been observed? Because, the answer goes, the bacteria that incorporate foreign genes are at a disadvantage and fail to survive. "That is an *ad hoc* argument," retorts Sinsheimer. "it is even worse than *ad hoc*, it is contrived."

Two cases are known in which bacteria seem to produce mammalian-type proteins. One bacterium produces a protein with some aspects of the activity of a reproductive hormone (known as chorionic gonadotropin). A class of bacteria make proteins similar to one of the digestive proteins produced in the mammalian stomach. Did these bacteria acquire mammalian genes to make their proteins? Or have they produced mammalian proteins by a process of convergent evolution? It's too early to say.

A general argument raised against the barrier theorem is that all the genetic combinations which can occur have already occurred in the course of evolution. Sinsheimer feels intuitively that this is not the case. He observes too that many who make the argument also speak of the benefits of genetic engineering—a proposition which is predicated on the opposite assumption, that beneficial organisms can be made which have not yet occurred in evolution.

Sinsheimer first put forward his warning about the procaryote-eucaryote barrier in February 1976, during the NIH's review of the final guidelines. No change was made in the guidelines as a result of Sinsheimer's warning, but his argument was not refuted either. "The fact is that we do not know which of the above-stated propositions [Sinsheimer's or a counterargument] is correct," the NIH director stated in publishing the guidelines.[16]

The possibility that gene splicing may be deliberately misused is another issue which Sinsheimer feels the NIH has neglected. The guidelines deal copiously with the health hazards to researchers but not at all with the possible danger to other sectors of society—with the fact, for example, that the technique is available for use not just by scientists but by entrepreneurs, horticulturists, the military, and terrorists.

"It may well be that there are some technologies that you should not use, not because they can't work, but because of the social dangers involved and the repression that would be necessary to prevent social danger," Sinsheimer remarks.

"We have gone along for several hundred years with the belief that knowledge and the means for acquiring knowledge are always beneficial. The situation that first led anybody to question that assumption was the atomic bomb. I think that a lot of people wish there were a way to forget all about nuclear physics, but there is not.

"For a while, many people hoped that was an anomaly. But now here comes another one. How do you cope with this new observation that some kinds of knowledge can be very dangerous? We have no assurances that science will not lead us into a very dangerous world.

"How do you control that without interfering with a lot of the freedoms that scientists have cherished? That is something we are only groping toward.

"How do you make policies for an issue which may take 50 years to resolve? Our government, at least in the past, has not been ready to make long-term decisions.

"Some of my colleagues feel that it is the scientist's job to do science, and society's job to cope with what he does. I disagree with this in principle. The scientist must keep the public informed because nobody else will.

"It is entirely possible, as Chargaff said, that the future may curse us. Really only the interests and concerns of the scientific community were involved in formulating the guidelines."

Yet, like other scientists, Sinsheimer is reluctant to

forego the gains in knowledge the new technique will bring. He does not believe that gene splicing should be banned but would prefer to see it confined, at least initially, to a handful of high containment facilities. In his view it does not make sense for local authorities to insist on safeguards stricter than those required nationally, and he has not used his position as chairman of the biology division at Caltech to prevent his colleagues from undertaking gene-splicing experiments.

Sinsheimer believes that one step leads inevitably to another, and that the gene-splicing techniques of today are the beginning of the genetic engineering of bacteria, of plants and domestic animals, and ultimately of man. "Do we want to assume the responsibility for life on this planet . . . ? Shall we take into our own hands our own future evolution?" Sinsheimer has asked. His answer is that we aren't clever enough, so shouldn't yet try.

NOTES

1. Testimony before the Senate Health Subcommittee, 22 September 1976.
2. Erwin Chargaff, "On the Dangers of Genetic Meddling," *Science* 192 (4 June 1976) 938.
3. Berg to Mayor Vellucci and the Cambridge City Council, 2 July 1976.
4. Federation of American Scientists, Press release, 23 June 1976.
5. *Boston Globe*, 28 and 29 March 1977.
6. Erwin Chargaff, "A Fever of Reason," *Annual Review of Biochemistry* 44 (1975).
7. Erwin Chargaff, "Profitable Wonders," *The Sciences*, August-September 1975, p. 21.
8. Chargaff, "On the Dangers of Genetic Meddling."
9. Robert Sinsheimer, "Troubled Dawn for Genetic Engineering," *New Scientist*, 16 October 1975, p. 148.
10. Robert Sinsheimer, "Genetic Engineering: The Modification of Man," *Impact of Science on Society* 20 (1970).
11. Robert Sinsheimer, "On Coupling Inquiry and Wisdom," unpublished lecture, June 1976.
12. This and following otherwise unattributed quotations taken from my article, "Recombinant DNA: A Critic Questions the Right to Free Inquiry," *Science* 194 (15 October 1976): 303.

13. Robert Sinsheimer, "An Evolutionary Perspective for Genetic Engineering," *New Scientist*, 20 January 1977, p. 150.
14. Ibid.
15. Sinsheimer to Donald Fredrickson, 5 February 1976.
16. National Institutes of Health, *Decision of the Director, National Institutes of Health, To Release Guidelines for Research on Recombinant DNA Molecules*, 23 June 1976, p. 8. ˙

10

GENE SPLICERS' PROMISES

" 'RECOMBINANT DNA technology' has the potential to make modern biology more relevant to the problems of human disease. This new way of studying genes simplifies complex biological problems so their individual parts can be analyzed. A deeper understanding of how the systems in human beings function is sure to come from these new methods. From this new understanding, novel methods are likely to evolve for preventing and treating the presently unconquered diseases that afflict us. Recombinant DNA research is our best hope for understanding diseases like cancer, heart disease, and malfunctions of the immune system, for which the prospects are poor for prevention solely by public health measures."

So David Baltimore of MIT told Senator Edward Kennedy at a hearing before the Senate health subcommittee in September 1976. Halsted Holman of Stanford University, on the other hand, offered the Senate committee a different view:

"It is by no means clear that the heralded medical and agricultural advances from recombinant DNA technology will accrue. They are only possibilities. Each of the examples given for ameliorating disease or improving

115

agricultural yield may be accomplished by other methods now being developed without analogous risk. Here too the experience with nuclear energy is illustrative. The promise of great benefit was the cornerstone of the defense of nuclear experimentation; it is now clear thirty years later that realizing that promise is at least problematical."

Despite the divergence of emphasis, there is merit in both views. The advances in knowledge brought by gene splicing, and the fact that it is a synthetic as well as an analytic technique, will inevitably improve the chances of devising fundamental treatments for some diseases.

It may be adventurous, however, to name any particular disease as candidate for a recombinant DNA cure. The only benefit that can yet be predicted with certainty is an increase in knowledge, and basic understanding of disease is usually but not always a sure road to treatment. The fundamental cause of sickle-cell anemia—the change of a single amino-acid unit in the hemoglobin molecule— has been known for years without the knowledge having led to a cure.

Biologists would wish to pursue gene splicing just as vehemently if it offered no foreseeable benefit to man or beast. The promise of benefits has been stressed in order to counter the arguments of Chargaff and Sinsheimer and to sway the public, whom, it was feared, would neither appreciate the value of pure research nor be prepared to accept even a minimum of risk for the sake of it. Yet if the proponents of the research have sometimes overpromised a little, or made the probable benefits seem just a little more tangible than they really are, the opponents have perhaps been overconfident in claiming that alternative techniques would do almost everything that can be done by gene splicing.

In parallel with the dispute about the benefits to be expected from gene splicing there has developed another argument—that of freedom of inquiry. Sinsheimer first raised the issue by saying that it won't do any more to "wave the flag of Galileo," meaning that scientists should not claim the absolute right to inquire into whatever pleases them, regardless of the risks to society. Proponents of gene splicing have reversed the argument and warned that even the attempts to regulate research may infringe academic freedom if taken too far. To which the opponents counter that gene splicing is just a technique that needs to be controlled, not a truth that anyone is trying to suppress.

The gene-splicing technique was designed by research scientists, and it is first and foremost a marvelous tool for analyzing the hereditary material. It offers a handle on problems that would have been arduous or impossible to solve with other present biological techniques.

Hitherto the chief barrier to understanding living cells has been the irreducible complexity of their chemical processes and regulatory systems. With restriction enzymes precise parts of the mechanism can be disected out piece by piece, and each part may then be amplified by the gene-splicing technique. Understanding the entire genetic program by which an organism develops from egg to adult now no longer seems so distant a goal.

The first fruits of gene splicing will be a better understanding of how cells work, of the fundamental differences between procaryotes and eucaryotes, and perhaps of the basic processes and history of evolution. It would be mistaken to assume that all secrets will be yielded up to the new technique. Much of what happened in evolution may be no more susceptible of reconstruction than Cleopatra's nose or the dust of Alexander. Even when the sequence of the human gene set has been determined in as much detail as anyone wants to know, we may still understand no more about human evolution and development than we now do about MS2 and phi-X174, two small viruses whose complete gene set has recently been determined.

Nor, despite the power of the technique, is nature likely to unveil herself in any great haste. Gene-splicing experiments have been in progress for more than two years, and some interesting facts have been gathered, but data collection is only a preliminary to understanding. Francis Crick summarized the state of the gene-splicing art at a conference held in March 1977: "People have . . . some results, but no deep results. We are in an era of very rapid progress. But it will be two to three years, maybe even just one year, before we see the sunlight, and even more rapid progress."[1]

The new technique is a synthetic as well as an analytic tool, and the range of practical possibilities it may open up is almost as wide as the imagination. Two general types of application can be envisaged. One is the redesign of living organisms, but most such applications depend on the acquisition of a lot more basic knowledge and therefore lie perhaps a decade or so in the future. Much nearer at hand is a second use of gene splicing, that of programming bacteria to mass-produce useful biological substances. The essence of the method is to construct or isolate the gene which specifies the desired product, insert the gene into bacteria so that the bacterial cells obey the new gene's instructions as they would that of their own genes; and then culture the bacteria and harvest for the gene product.

Just such a project is already underway for the production of human insulin. Present supplies come from cattle and pigs, which produce an insulin molecule sufficiently similar to the human version for therapeutic value, but the supply threatens to lag behind the increasing demand. No one yet knows how to pick out the insulin gene from among the million or so genes in each human cell. But because the chemical structure of the insulin protein is known, and because of the fixed chemical relationship between a protein and the gene by which it is specified, the chemical structure of the human gene for insulin (or more precisely, for the insulin precursor protein) can be deduced and synthesized.

A company called Genentech has contracted with researchers at Caltech to synthesize the human insulin precursor gene and with Herbert Boyer and colleagues at the University of California, San Francisco, to insert the gene into bacteria for cloning. The project cannot yet be guaranteed success, but it is at least close enough to feasibility to have attracted investment.

The same method may work for the mass production of any desired therapeutic protein, and if the economics were right, conceivably for edible proteins as well. Besides insulin, another early candidate for gene-splice production is likely to be interferon, the body's natural antiviral substance, which cannot be obtained in quantities sufficient even to conduct therapeutic tests. The human interferon gene, like the insulin gene, is still inaccessible, but scientists have isolated the next best thing, the chemical copy of the gene (known as messenger-RNA) that serves as the intermediary between the gene and its protein product. In the next few years it may be possible to use the interferon messenger-RNA as the basis for programming bacteria to produce useful quantities of the scarce protein. Similar methods might serve to produce the rare blood-clotting factor lacked by hemophiliacs as well as the antibodies to various disease agents and other therapeutically useful proteins.

The genes of viruses are much more accessible than those of cells because viruses generally possess only a handful of genes. For this reason, one of the earliest benefits of gene splicing may be making better vaccines. Viral vaccines are at present produced by culturing the viruses in living cells. When the cells are harvested to make a vaccine, other viruses or components of the cell often contaminate the preparation and cause allergic reactions or other side effects in the vaccinees. With the gene-splicing technique it should prove possible to isolate the gene for one specific component of the virus that will immunize the body as efficiently as would the whole virus itself. Bacteria programmed with the gene would then produce a pure protein product to be used as the vaccine.

The approach is already being used to develop a vaccine against a fatal enteric disease of livestock. The disease is caused by a toxin which is produced by a plasmid of *E. coli.* Researchers have succeeded in separating the toxin gene from the rest of the plasmid. The next step is to snip away parts of the DNA until the gene sequence that is left specifies a protein which is harmless but similar enough to the toxin to stimulate the animal's immune system into producing antibodies that attack the toxin.[2]

Industrial applications of gene splicing are unlikely to be confined to medical uses. It may be possible to tailor bacteria to synthesize useful products, such as methane gas, out of sewage sludge. Certain algae are known to produce hydrogen from water, using sunlight as their source of energy. With the help of gene-splicing techniques, the algae can perhaps be adapted to provide an economic source of pollution-free energy.

Genetically rejigging bacteria and algae in various ways is likely to produce the first practical fruits of gene splicing. Further ahead lies a different category of manipulation, the genetic modification of animals and higher plants. Plants are likely to be the first subject of this art because whole plants can be generated from a single cell of the adult organism, a plasticity animal cells do not naturally possess. Valuable new crop varieties might be produced from cells altered so as to produce better protein or use sunlight more efficiently.

The possibility of equipping crop plants with their own nitrogen-fixing genes is a frequently discussed application of the technique. The benefits would certainly be appreciable. The energy required to manufacture the world's supply of nitrogen fertilizer is equivalent to about two million barrels of oil a day. In addition, the farmer bears the cost of applying the fertilizer and the environment the cost of the pollution from nitrogen runoff. The only organisms at present known to "fix" the nitrogen in the atmosphere—in other words, to turn the gas into a chemical form usable by plants—are certain bacteria that possess the nitrogen-fixation genes. Leguminous plants, such as peas and soybeans, harbor colonies of these bacteria in nodules on their roots. No one knows how to make nonleguminous crops, such as wheat or corn, learn to produce similar nodules, but a simpler trick might be to introduce the nitrogen-fixing genes directly into the plant's own cells, removing the need for nitrogen fertilizer.

While well worth trying, the idea may prove totally impractical. Leguminous plants seem to devote a large fraction of their available energy budget to the support of their bacterial guests, a burden that other crop plants might not be able to absorb. Another problem is that the nitrogen-fixing enzyme is poisoned by oxygen, a substance that plants produce.[3]

The genetic engineering of animals seems a considerably more distant prospect even than that of tailoring plants to man's design. Two different kinds of animal engineering can be envisaged. First, there is treatment applied to particular organs of the whole animal, such as to defective bone marrow. The effect of such a manipulation would be coterminous with the individual's life. A more permanent kind of engineering would be alterations to the organism's germ-line cells, which would be perpetuated in its descendants. The difference is between genetically treating an individual and genetically engineering a species.

Genetic treatment of human diseases is a concept so

often discussed as to have acquired a name—gene therapy. In some genetic diseases a specific gene is defective, or absent; in others the fault may lie with the regulatory mechanisms whereby a gene is switched on or off. If it becomes possible to isolate the normal version of the functional or regulatory genes which are deficient in blood disorders, such as thalassemia or sickle-cell anemia, conceivably the normal genes could be spliced onto a plasmid or virus, which would then be introduced into the bone marrow cells of people suffering from the disease. Any such scheme sounds farfetched at present, but equally it would be unreasonable to assert that manipulations of this general kind will never be possible.

Skeptics who argue that the promised benefits of gene splicing are far distant, illusory, or attainable by other methods, will doubtless prove correct in at least some of the instances now being discussed. But production of better vaccines is one application that seems already to lie within grasp. Some critics have disputed the benefits alleged for curing cancer on the grounds that the disease can better be reduced through known public health measures, such as removing carcinogens from the environment. Similarly for cholera the argument goes, Why do we need gene splicing to make an anticholera vaccine when proper sanitation would be surer, safer, and cheaper? The answer to such arguments is, Why not use both approaches?

Yet the critics may have a point of some relevance in arguing that any medical advance to come from gene splicing will be limited by the efficiency of the health delivery system. Ethan Signer of MIT puts it this way:

> What's more, however we get them, having these benefits probably won't make much of a difference. Right now we have a lot that is potentially available, yet does not reach the public at all effectively. . . . Right now even the rich have a low standard of medical care, and the poor of course a much lower one. Any benefits of recombinant

DNA will fit right into this pattern. Miracle cures won't suddenly find their way to rural or ghetto hospitals, and drug companies won't suddenly put health before profit. We *are* stalled in medical care, but not for lack of recombinant DNA. And all this focus on recombinant DNA is going to draw even more resources away from what we really need. If we're worried about our people's health, let our main course of action be to give them what we can already. Then we can have the pie in the sky for dessert. . . .

We can and should justify basic research entirely on its own merits. And I believe we can trust the public to support it on those grounds. But we ought to stop peddling miracle cures and sniping with magic bullets. Recombinant DNA isn't going to cure what ails you, although good medical care might. The anti-science backlash is real, and we're going to lose the public if we're not more candid about what research can and cannot promise, and about why we want to do it.[4]

Signer has decided not to use gene splicing in his research, although it would have been ideal for certain experiments he wished to perform. The same decision has been made by Jon Beckwith of the Harvard Medical School. "I do not wish to contribute to the development of a technology which I believe will have profound and harmful effects on this society," Beckwith explained to a National Academy forum held in March 1977. It was not that he believed the technology of human genetic engineering to be inherently dangerous; rather, he didn't trust society to make proper use of it.

Signer's and Beckwith's decisions are acts of principle that may well handicap their research and scientific careers in one way or another. But voluntary acts of self-denial are quite different from bans imposed by outsiders, and that is what some scientists claim that public bodies are coming close to doing in their attempts to regulate gene splicing. The first issue of a periodical devoted to recombinant DNA methods warned that science was facing a threat similar to the suppression of biology in Lysenko's Russia and of astronomy during the persecution of Galileo.

Such alarms fail to make the distinction between suppressing a theory or idea and regulating a technique so as to ensure no inadvertent harm is caused to the public. So far at least, actions by Congress and local authorities to control recombinant DNA research have no more been attempts to suppress biologists' freedom of thought than are radiation standards an attempt to anathematize nuclear physics.

Yet proponents of the research are probably correct in divining that fear of the ultimate uses of gene splicing is as much a factor in legislators' itch to regulate the research as is any apprehension about the immediate health hazards. The proponents' promises of gene therapy and cracking the structure of the human gene set have probably contributed to this concern about ultimate uses just as much as the opponents' lurid warnings of malign applications in the style of *Brave New World.*

To lay to rest these specters the defenders of the research have emphasized the distinction between knowledge and the uses that are made of it. "In order to counteract the growing pessimism about the nature of knowledge, the proper separation of science from technology must be made and, in the continuing dialogue, the distinct values and problems in each must be carefully articulated," Maxine Singer of the NIH suggested at the National Academy forum. In similar vein, Stanley Cohen urged those concerned about the appropriate applications of knowledge "to address themselves to the real issue, which is what is done with knowledge by society, and not to the knowledge itself."

"Recombinant DNA is not a search for truth, it is not a search for knowledge, it is only a technique," was Ethan Signer's response. Or as his MIT colleague Jonathan King chose to put it, "None of us is saying, 'Don't accumulate knowledge.' What if we cut off this particular technology—will the font of knowledge dry up? For Christ's sake, no."

The font of knowledge would not dry up, yet to lop off any branch of research without excellent reason, or even to ban a mere research tool arbitrarily, would make a chilling precedent. Yet however much the proponents might hawk the benfits and the opponents harp on the risks, the estimation of both has involved social values as well as scientific facts, and social values are a matter for society to decide.

NOTES

1. Quoted in *Science News,* 2 April 1977, p. 217.
2. Stanley Cohen, "Recombinant DNA: Fact and Fiction," *Science* 195 (18 February 1977): 656.
3. K. T. Shanmugam and Raymond C. Valentine, "Molecular Biology of Nitrogen Fixation," *Science* 187 (14 March 1975): 919.
4. Ethan Signer, "Recombinant DNA Will Not Cure What Ails You" (Testimony before the House health subcommittee, 15 March 1977).

11

FROM CAMPUS
TO CONGRESS
VIA CITY HALL

Since scientists had taken the initiative to control the new technique, it was only reasonable for them to hope that they would be allowed to regulate themselves. The desire to have their own patron agency, the National Institutes of Health, control gene splicing on an informal basis was not just a matter of convenience but stemmed from apprehension that the public, should it once get involved in the issue, would demonstrate its assumed antipathy to science by crippling the research or shutting it down altogether.

Both the hope and the fear were to prove overdrawn. The events of June 1976 showed that gene splicing was too hot an issue for scientists to keep to themselves. Yet the public, once it was given a chance to express an opinion on the matter, turned out to be concerned about the research but not hostile.

When the NIH committee completed its guidelines on recombinant DNA research, the NIH decided to apply them on a voluntary basis rather than issuing them as government regulations. But the voluntary approach soon ran into trouble for the obvious reason that it only had bite for scientists who were funded by the NIH, and whose support could be cut off if they violated the guidelines. The agency tried to remedy the deficiency by persuading all other patrons of research to apply the NIH guidelines to their scientists as well. Government agencies, such as the National Science Foundation, agreed to do so, and the Pharmaceutical Manufacturers Association (PMA), speaking for the major drug companies, said its members would adhere to the NIH rules provided that they could be exempted from the requirement to make public their research plans.

But that still left outside the NIH's corral all researchers funded by private foundations and by companies who were not members of the PMA, ranging from General Electric to the Cetus Corporation. The failure of the guidelines to apply to everyone made the NIH's approach unacceptable to public interest groups and state legislatures.

Another factor which made federal legislation necessary was that local authorities, starting with the City Council of Cambridge, Massachusetts, set about framing their own regulations on gene splicing. To prevent a crazy-quilt pattern of different laws throughout the country, the NIH decided in early 1977 to push for a federal law that would make their guidelines applicable to everyone and at the same time strike down or preempt the state and local laws. Hearings were held by the House and Senate health subcommittees in March and April 1977, during which it became clear that the NIH guidelines were acceptable as a basis for regulation. The federal law (likely to be enacted in summer 1977) will probably make the NIH guidelines mandatory for all researchers, whether academic or industrial, and will institute some form of licencing.

The celebrated hearings before the Cambridge City Council at which Mayor Alfred Vellucci put Harvard and MIT scientists on the griddle were of less lasting interest than what happened before and after. The roots of the imbroglio at Cambridge lay in the proposal to build a P3-style facility on the fourth floor of Harvard's Biological Laboratories. The original purpose of the facility was to culture animal cells, but when it was decided to conduct gene-splicing experiments there as well, a heated debate broke out among researchers in the building and soon spilled over into the university at large.

The issue divided many scientists who had been on the same side of various liberal or left-wing causes. One of the sponsors of the lab, Mark Ptashne, had gone on a lecture tour to North Vietnam in the winter of 1970. Another sponsor, biology department chairman Matthew Meselson, has conducted a lengthy campaign against chemical and biological warfare, and is credited with a major role in the closing down of Fort Detrick and the renunciation by the United States of offensive biological warfare in 1969. David Baltimore, MIT's leading defender of the research, was active in opposition to the Vietnam War.

Ranged on the other side of the Cambridge gene-splicing debate were such left-wing luminaries as George Wald, winner of a Nobel Prize for his work on how the eye sees color, who had employed his gift of oratory against the Vietnam War and more recently in other causes, such as opposition to nuclear power; also Ethan Signer of MIT, one of the first scientists to visit China; and Jon Beckwith of the Harvard Medical School, who had publicly warned of military misuse of science when he isolated a pure gene in 1969.

Biologist Ruth Hubbard was also prominent among the Cambridge critics—she and husband Wald work in the same building as the planned P3 facility—as were Jonathan King of MIT, a member of the radical group known as Science for the People, and Richard Goldstein of the Harvard Medical School, an active participant in the Boston Area Recombinant DNA Group.

The issue of the P3 lab was vigorously argued at a series of faculty committee meetings, but nowhere more intensely than among researchers in the Bio Labs, who felt themselves directly at risk. The building is infested with a seemingly ineradicable insect pest, known as the pharaoh ant, which, it was feared, would spread organisms from the P3 facility around the building. Electrical failures and the frequent floods from broken pipes would also assist the dispersal of gene-spliced organisms, the critics pointed out, in addition to which the building was located in a heavily populated area; in short, they said, it was probably the worst possible site that could have been chosen as the home of a P3 facility—so why not build the lab someplace else?

The proponents' answer, in essence, was that the risk was too slight to justify the inconvenience. As Ptashne explained to Harvard's Committee on Research Policy at a meeting on May 28, 1976:

> There would be pressure to take things out of the lab and bring them back into the lab if the lab were located at a distance. And there is the fact of overwhelming inconvenience: A P3 lab is not a lab you live in. It is a lab you go to and do an experiment in and walk out of. You never sit down in it. To ask somebody to walk from the Bio Lab to the Accelerator, for example, and walk back ten times a day to do an experiment makes the lab useless.[1]

Harvard was so engrossed in its internal debate about the plans to build a P3 lab that it forgot to mention them to the mayor of Cambridge. Alfred Vellucci read about them in a well-researched article[2] in the *Boston Phoenix*.

While he was discussing the article with his aides, he was told that George Wald and Ruth Hubbard had come to see him. They described the dangers of the research and told Vellucci they would be willing to testify at a meeting of the city council. The mayor got his council to agree by a 9-to-0 vote to hold public hearings. "We want to be damned sure the people in Cambridge won't be affected by anything that would crawl out of that laboratory," said Vellucci in announcing the hearing.⌋

The move aroused instant consternation throughout the scientific establishment. A deluge of letters descended on the Cambridge City Hall from Nobel Prize winners under the impression that their word would influence the mayor to relent. All that Vellucci cared about Nobel Prize winners was that he had one, Wald, on his side. Nor was he moved by correspondents who feared that the council was "considering suppression of a serious and responsible search for new knowledge." Vellucci does not take new knowledge on trust. As he explained to a reporter, "It's about time the scientists began to throw all their goddamned shit right out on the table so that we can discuss it. . . . Who the hell do the scientists think that they are that they can take federal tax dollars that are coming out of our tax returns and do research work that we then cannot come in and question?"[3]

During his quarter century in city politics, Vellucci has often harried the academic institutions within his domain, but often with a serious reason beneath the posturing. His celebrated threat to pave over Harvard Yard for a parking lot did achieve its purpose, according to Vellucci—the university built more student parking facilities to relieve congestion in the city streets.

On the gene-splicing issue, too, there seemed to be a measure of serious concern beneath the colorful talk about monsters crawling out the Harvard laboratories.

The days of public hearings held by the City Council on June 23 and July 7, 1976, produced lots of press and TV coverage for Vellucci's hectoring of the Harvard scientists and threats to close down research. Yet the mayor overplayed his strength with his council. When the vote was taken, he was turned down on his resolution to ban all gene-splicing research from Cambridge for two years. Instead, the council approved by 5 to 4 a resolution by Councillor David Clem calling for a three-month "good faith" moratorium applying only to research requiring P3 or P4 facilities.

At the same time the council voted to set up a review board to decide whether P3 research—there are no plans to do P4 work—should proceed and, if so, under what conditions. The crucial task of selecting the membership of the board was entrusted not to the mayor but to the city manager, James Sullivan. Deciding that scientists were hopelessly divided about the issue and likely to stay that way, Sullivan appointed a panel of citizens, none of whom, apart from a doctor, had any familiarity with biological research. The chairman of the board was Daniel Hayes, a former mayor and owner of a heating oil business in Cambridge. Other members were Mary Nicoloro, a community worker (and cousin of Vellucci's); Sister Mary Lucille Banach, a hospital nurse; Sheldon Krimsky, a professor of urban policy at Tufts University; William Le Messurier, a structural engineer; Cornelia Wheeler, a former city councillor; John Brusch, a physician specializing in infectious diseases; and Constance Hughes, a nurse and social worker.

Contrary to some expectations, the citizens took the time—some seventy-five hours of hearings—to acquire an informed understanding of the issue. Their decision, announced on January 5, 1977, had an importance that extended far beyond Cambridge because it represented the first time that the public had been given the opportunity to declare its own unaided opinion on the risks and benefits of the gene-splicing technique.[4]

The citizens chose their own ground for decision, without falling captive to either side of the scientists' debate. The proponents had implied that restricting the research would impede discovery of a cure for cancer and the like, but the review board decided that "the benefits to be derived from this research are uncertain at this time," although the possibility for advancement certainly existed.

The opponents had said that since no containment system could be foolproof, the research should not take place in a city, if at all. The review board decided that absolute assurance of safety was an unreasonable expectation. The citizens did not define precisely what degree of risk was acceptable, but they at least grasped the metal that had hitherto been too hot for any other group to handle by deciding that there was a risk but that they, on the public's behalf, were willing to accept it even without any immediate countervailing benefit.

"Knowledge, whether for its own sake or for its potential benefits to mankind, cannot serve as a justification for introducing risks to the public unless an informed citizenry is willing to accept those risks," the review board wrote in its report. P3 research, it advised, should go ahead, although under certain additional safety conditions to those specified in the NIH guidelines, such as proper monitoring for the escape of the organisms used in the experiments.

The citizens' recommendations were accepted by the city council in February 1977, although with a few further restrictions, such as a ban on P4 research, which the universities have no known plans to perform. Biologists at Harvard and MIT now have to work under more stringent conditions than if the city had left them alone, but the extra conditions are tolerable and confer the advantage of enabling the research to proceed with the informed consent and approval of the public.

In retrospect, at least, scientists at Cambridge and elsewhere probably overreacted to Vellucci's threats to drive gene-splicing research out of the city. Fear of what

Vellucci might do spilled over into antagonism against Wald, Hubbard, King, and others who gave evidence against the P3 lab at the city council hearings. In the tense atmosphere within Harvard younger scientists began to mute their criticisms for fear of jeopardizing their academic careers.

Harvard biologist Ursula Goodenough says she dropped out of the debate at an early stage: "I decided that, while I was happy to speak out on things that really concern me, I didn't want to lay my head on this particular block at this stage in my career. A lot of people dropped out at the same time I did, probably for the same reasons. All of us were feeling very pressured, but it was very subtle stuff. Those of us who aren't tenured and who know how difficult it will be to keep our jobs get anxious when there are bad feelings between us and our tenured colleagues. It is unwritten things like a hostile reaction in the hallway."

Goodenough's colleague Ruth Hubbard attributes the hostility to the fact that the opponents offended academic mores by taking their criticisms to a public forum: "That is the sin for which there is no forgiveness, not that we disagree on a scientific issue" Hubbard remarks. MIT historian of science Rae Goodell agrees: "There are rigid rules about what a scientist should and should not do. It's fine to be critical in private but not in public. If you want to express social responsibility, it is fine to do so in Washington but not on the street. Pressure on the critics in this issue would seem absolutely inevitable."

Even the citizens review board noticed the animus toward the critics and thanked them in its report for giving testimony. "If you have criticized those who might be in a position to affect your advancement in an academic field, you are putting yourself on the line. I think it was very courageous for them to come forward. They put themselves open to future harassment from their fellow scientists," comments review board chairman Daniel Hayes.

Harassment of the Cambridge critics seems to have been mostly verbal and has not prevented their side of the case from reaching the public record. Yet, as was observed in an NIH statement on gene splicing, "a key element in achieving and maintaining . . . public trust is for the scientific community to ensure an openness and candor in its proceedings."[5] The hostility at Harvard and MIT toward the critics of gene splicing is the one shadow on what has otherwise been a notably open and candid process.

The Cambridge citizens review board may well be judged to have proved its belief that "a predominantly lay citizen group can face a technical scientific matter of general and deep public concern, educate itself appropriately to the task, and reach a fair decision." Yet even the review board, like all its predecessors, focused on the immediate public health hazards, viewing the longer range issues of gene splicing as beyond its scope.

Hayes's own belief is that one should strive for the goal of conquering all human disease and face any risks as they materialize, rather than holding back for fear of hypothetical dangers. Another perspective is that of Councillor Clem, who played a major role in bringing about the eventual compromise: "I have a gut feeling that ten to fifteen years from now I am going to regret having worked toward a compromise on this issue, because I think we are stretching our limits of being able to respond in a civilized way to the fruits of knowledge. We are becoming fat with all this knowledge, so fat and bloated we may not survive. One of the things that my generation is mindful of is that in the 1960s we should have listened to the dissenters on Vietnam, on the despoliation of the environment. Sometimes I worry about being in a position of trying to compromise until you compromise too far."

The events at Cambridge sparked off similar dialogues between city hall and campus in cities from Bloomington, Indiana, to San Diego, California, indicating a remarkable degree of public interest in the issue. All communities have accepted the NIH guidelines as a basis for control, but most have suggested further requirements, generally in the form of strengthening the enforcement of the guidelines:

New York State: The state attorney general's Environmental Health Bureau held hearings in October 1976 and has prepared a bill to regulate all gene-splicing research in the state. "Unregulated recombinant DNA activities," in the view of a report prepared by the bureau, "pose a unique threat to the public health and the environment."[6] The bureau recommended that all users of the technique, whether in universities or in industry, should be required to obtain a license from the State Department of Health and that all laboratories should set up programs of medical surveillance to check on infections of their personnel.

California: A tussle has developed in the state legislature between the State Department of Health, which drafted a stringent bill to regulate all hazardous biological research, and the Assembly Committee on Health, which has been working with the universities to draw the teeth from the bill, first introduced in the legislature on March 3, 1977. There is some sentiment for writing in a self-destruct clause that would nullify the bill if federal legislation is passed. Both sides have been vying for the ear of Governor Jerry Brown, who was reportedly turned on to the hazards of recombinant DNA research by singer Linda Rondstadt. Brown received a lengthy briefing from James Watson of Cold Spring Harbor and Halsted Holman of Stanford, representing the proponents and opponents respectively, but he has not yet announced his position.

Maryland: The Maryland General Assembly on April 11, 1977, enacted the first state legislation in the country to extend the NIH guidelines to private industry. Companies doing recombinant DNA research must register their activities with the state and set up a local biohazards committee with citizens included in the membership. The Maryland law will go out of effect if a federal law on gene splicing is enacted.

New Jersey: State Attorney General William F. Hyland, whose interest in biomedical issues was demonstrated during his handling of the Karen Quinlan case, has been following the gene-splicing issue closely. His assistant on the subject, Dennis Helms, says his own feeling —Hyland has not yet come to a decision—is that state regulation is not a good idea for an issue that can properly be settled only on a national basis. There is no point in driving the research underground by excessive regulation, Helms believes, because "in the end we are going to depend on the responsibility of the individual scientist. But I can assure you the response will be electrifying if there is a bad accident. That will mean banning everything in the ridiculous fashion that always happens when you do things too fast."

San Diego: Seeking to avoid a Cambridge-style confrontation, the University of California at San Diego informed Mayor Pete Wilson in July 1976 of its intention to build two P3 facilities. The mayor asked his Quality of Life Board to set up a DNA Study Committee chaired by Albert Johnson, dean of sciences at the university.

After hearing witnesses from both sides, the committee recommended in a report of January 26, 1977, that the NIH guidelines be endorsed but with several further requirements, such as keeping proper health records of experimenters.[7] The desirability of confining all recombinant DNA research to P3 facilities should be considered, the committee advised.

The committee, whose membership was dominated by public health experts, also noted certain "weaknesses" in the NIH guidelines, such as the lack of discussion of the ecological impacts of gene-splicing research: "There is almost no evidence that they have been considered seriously," the committee noted, adding that in the event of an accident, the lower the position on a food chain of the escaped organism, the greater the potential for ecological disruption.

The San Diego committee also alluded to the potential conflict of interest in the NIH's position as both the supporter and regulator of gene-splicing research: "Eventually it may be necessary to divide the research and regulatory functions in a manner similar to the way that nuclear research and regulation have recently been separated."

Madison, Wisconsin: The mayor's office in December 1976 prepared a resolution that would set up a citizens committee and arrange for a public debate on the issue. The University of Wisconsin instituted a committee at the same time, possibly with the idea of heading off a citizens inquiry. "There is now a little bit of jostling going on between the city and the university about whether there should be a public debate," an aide to the mayor remarked in February 1977.

Princeton, New Jersey: A university committee chaired by ecologist Robert May recommended in a report of December 1976 that research should be permitted at Princeton under conditions somewhat stricter than those of the NIH guidelines.[8] But a public meeting organized jointly by the town and university resolved to appoint a citizens committee to review the report. The university decided to permit no recombinant DNA research until the citizens review is received.

Bloomington, Indiana: The mayor's office, watching the events in Cambridge, heard rumors of gene-splicing research in progress at the University of Indiana and was later told by the university of plans to build a P3 laboratory. The mayor's environmental commission held hearings but found no serious fault with the university's procedures. "For the most part the community is satisfied that the university is being responsible. We don't anticipate taking action at the present time," an aide to the mayor said in February 1977.

Ann Arbor, Michigan: Mayor Albert Wheeler, who happens to be a microbiologist, is taking no action. One reason may be that the University of Michigan has gone through a more intense debate about the technique than any other institution, Harvard and MIT included. Vigorous opposition to the research was mounted by Susan Wright, a historian of science at UM's college of engineering. Wright stirred up such a commotion that one of her supporters has said—and the proponents of the research agree—that the university will never be the same again.

In the belief that gene splicing was the wave of the future, the university decided in April 1975 to construct three P3 facilities. It also set up three committees to study various aspects of the research, one of which, committee B, was assigned to ponder the ethical and social implications of gene splicing. Composed of humanists as well as natural scientists, the majority of the committee concluded in March 1976 that the research should go ahead.[9] The lone dissenter, historian Shaw Livermore, based his objections not on the immediate hazard to health—in his opinion that risk was amply taken care of—but on the consequences of creating new forms of life. Livermore's brief statement said, in part:

> If such a research effort is successful man will have a dramatically powerful means of changing the order of life. I know of no more elemental capability, even including manipulation of nuclear forces. While it clearly would

present opportunities for meeting present sources of human distress, I believe that the limitations of our social capacities for directing such a capability to fulfilling human purposes will bring with it a train of awesome and possibly disastrous consequences. Decisions will be made by individuals, groups, and perhaps whole societies that may well have unintended but irreversible effects. . . .

Moral decisions cannot be tested by quantifiable, scientific or even strictly logical modes of proof. They do not consist of precise "cost-benefit" analyses, however elegant. . . . I find unpersuasive the notion that all temporarily safe means to relieve human distress are justifiable. The highest purposes of moral reflection must be to keep the terms of human life not only tolerable but genuinely fulfilling. I do not sense that we are presently at such a crisis, or are so powerless, that we must suspend judgment to grasp at each prospect of temporary alleviation. If this requires that humankind and science move towards a new understanding at some limits then we must begin. That this understanding will be an uneasy one makes it no different from most important human arrangements.

Far worse are the barely heard sounds of quiet desperation.

In an interesting response to Livermore's dissent, committee member Robert Burt, a lawyer, observed that value perspectives about various technologies tend to change over time: "We frequently find it hard to understand why preceding generations were so exercised about these technologies, and our contemporary puzzlement suggests the parochiality of both past and current views about technology uses." Hence, Burt said, recombinant DNA research should be allowed to proceed until it became quite clear that the technology derived from it was leading in abhorrent directions.

The University of Michigan's board of regents held five meetings on the matter, deciding in May 1976 by a 6-to-1 vote to let the research proceed. Robert Helling, head of one of the two UM research groups using the technique and a member of the four-person team that invented it, says that the debate took up an enormous amount of time—he was unable to do any research for about a year—but that the argument had been well mannered. "There have been intense feelings at times but there has never been anything personal, we have tried to keep things civilized," Helling reflects.

No less involved than the academic community, but with its debates held behind closed doors, has been the pharmaceutical industry and other companies with an interest in putting the gene splicer's art to practical effect. The positions of the drug companies are coordinated by their Washington trade association, the Pharmaceutical Manufacturers Association. The PMA's initial stance was that it would comply voluntarily with the NIH guidelines, provided that companies were exempted from the requirement to disclose their research activities. But by February 1977 the PMA had abandoned the voluntary approach. "We now realize that position was not pragmatic," said PMA scientific director John G. Adams: "We are in a fishbowl on this. There are charges that we are doing something clandestine." To protect itself from public suspicion the PMA decided to back proposals for a government registry of research to which companies would confide their gene-splicing activities. But the proposed registry, in the PMA's view, would have to be immune from public disclosure under the Freedom of Information Act in order to protect trade secrets.

As of mid-1977 only six major drug companies had an active interest in the new technique—Hoffman-LaRoche; Upjohn; Eli Lilly; Smith, Kline, and French; Merck; and Miles Laboratories. A smaller company with a keen eye on gene splicing is the Cetus Corporation of Berkeley, California. Cetus has as consultants Stanley Cohen, one of the inventors of the technique, and bacterial geneticist Joshua Lederberg, both of Stanford University. The company's present specialty is improving the genetics of industrial microorganisms, but gene splicing "will be a very major aspect of our future output," corporation president Ronald E. Cape has said.

A second small company with a head start in the field is Genentech, which has signed up another of the technique's pioneers, Herbert Boyer of the University of California at San Francisco. Boyer and colleagues are working on a project aimed at the commercial synthesis of human insulin and of another hormone, somatostatin.

On the basis of Cohen and Boyer's development of the art Stanford University and the University of California have applied for a patent on the commercial uses of gene splicing. The patent would apply only to manufacture, not to academic or industrial research uses of the technique.

Environmentalist organizations have been following the progress of gene splicing as closely as has industry, and several groups have prepared policy questions on the research. The Environmental Defense Fund and the Natural Resources Defense Council petitioned the Department of Health, Education, and Welfare for a public hearing to determine whether any gene-splicing research should be allowed and if so, under what conditions.[10] Such a hearing, the petition stated, would serve "as a broad-based public review of the existing NIH guidelines and would permit open debate on issues given little attention by the NIH drafting committee of the office of the director," such as whether *E. coli* is an appropriate experimental host.

At a meeting on January 9, 1977, the board of directors of the Sierra Club decided that, pending further information and discussion, "the Sierra Club opposes the creation of recombinant DNA for any purpose, save in a small number of maximum containment labs operated or controlled directly by the federal government."

Friends of the Earth (FOE) wants a moratorium on all recombinant DNA research to be imposed immediately pending further public investigation. FOE organizer Francine Simring has also put together the Coalition for Responsible Genetic Research, a group that claims four hundred sponsors, including Nobel Prize winners George Wald and MacFarlane Burnett. At a March 7, 1977, press conference calling for an "immediate, international moratorium" on all gene-splicing research, Wald called the technique "perhaps the biggest issue that has ever arisen in the history of science" and "the biggest break with nature that has occurred in human history."

One sector of the public that has been strangely loath to say its piece on gene splicing is the religious community. Creation by man of new forms of life is a subject on which the church might be expected to have an opinion. Gene-splicing research was indeed discussed in *The Pilot*, the official newspaper of the Archdiocese of Boston.[11] But the article, by Bishop Thomas J. Riley, studiously avoids the question of man's approaching power to create new Gardens of Eden in his backyard, at least on the microbial level. The problems of gene splicing, Riley concludes, "can definitely be settled only by scientists themselves."

A more reserved position has been taken by the Committee for Human Values of the National Conference of Catholic Bishops. In a statement of 4 May 1977, the bishops noted that science is not value-free, but often carries ethical implications which require reflection.[12]

"The Church, while recognizing its limitations in scientific matters, has something to contribute to this reflection," say the bishops. Their statement warns against using utilitarian values—weighing risks against benefits—as the only yardstick for judging gene-splicing. There are moral values to consider as well: "A good end or good purpose does not justify any means. There might well be a worthy scientific goal which ought not to be pursued if it unjustifiably violates another human good. In other

words, ethical constraints might slow down, or even preclude, some scientific advances." The bishops make no
present judgment about gene splicing but this, they say,
is the guideline which moral reasoning about the research
should follow.

NOTES

1. "Recombinant DNA Hearing at Harvard," *Boston University Journal* 24, 3 (1976).

2. Charles Gottlieb and Ross Jerome, "Biohazards at Harvard," *The Boston Phoenix*, 8 June 1976.

3. Arthur Lubow, "Playing God with DNA," *New Times*, January 1977.

4. Cambridge Experimentation Review Board, "Guidelines for the Use of Recombinant DNA Molecule Technology in the City of Cambridge" (submitted to the city manager, 5 January 1977).

5. National Institutes of Health, "Draft Environmental Impact Statement for Guidelines for Research Involving Recombinant DNA Molecules," 19 August 1976, p. 16.

6. Report and Recommendations of the New York State Attorney General on Recombinant DNA Research, 8 February 1977.

7. Report of the DNA Study Committee to the Mayor and City Council, 26 January 1977.

8. Recommendations for the Conduct of Research with Biohaz-ardous Materials at Princeton University, 6 December 1976.

9. Report of the University Committee to Recommend Policy for the Molecular Genetics and Oncology Program (Committee B), March 1976.

10. Petition of Environmental Defense Fund, Inc., and Natural Resources Defense Council, Inc., to the Secretary of Health, Educa-tion, and Welfare to Hold Hearings and Promulgate Regulations Under the Public Health Service Act Governing Recombinant DNA Activities, 11 November 1976.

11. Bishop Thomas J. Riley, "Will DNA Research Evoke Moral Concern?" *The Pilot*, 7 January, 1977.

12. Statement on Recombinant DNA Research, 4 May 1977. Na-tional Conference of Catholic Bishops, Washington, D.C.

12

THE DILEMMAS OF DEMIURGY

THE JUDGMENTS involved in assessing the future direction, promises, and hazards of gene splicing are not easy, and only to a few is the truth apparent on first inspection. Most of the issues that have been so ardently discussed over the last few years will in any case probably seem of minor significance a few years ahead.

Despite the yodeling about medical benefits, the most important contributions of the gene splicer's craft may well come from some unexpected quarter, such as in the adaption of microorganisms to convert sunlight into machine-usable forms of energy. Our worst medical afflictions may prove incurable, gene splicing or no. Most forms of cancer are probably environmental in origin, and their exact causes, when identified, may well prove as perfectly easy to eradicate as that of lung cancer. As for heart disease, a thorough cure of that would be almost tantamount to a cure for aging, a medical advance which the world at its present population could well do without.

The risks posed by gene-splicing experiments to human health and the environment can as yet only be guessed, not exactly estimated. By most reasonable standards the risks to health can be assessed as small to vanishing, and the rules set by the NIH will probably prove to be an adequate initial response. The rules reflect the self-interests of their framers—for example, in the lack of strong requirements for monitoring the escape of organisms—but seem nevertheless to fall within the general range of what a more disinterested group would have recommended.

As for the broader risk to the general environment, the NIH's rules will certainly reduce substantially the outward flow of gene-spliced organisms from their place of manufacture. The test of what risk they pose to the environment may lie several years ahead when hundreds of laboratories, and possibly factories as well, are in the gene-splicing business. By that time the rules may have been relaxed, either officially or unofficially, and the trickle of recombinant DNA released into the environment may become a flood. But by that time also, the hazards may have been better assessed.

Whatever the risk, there are two safety nets against serious harm. From increasing knowledge, or minor mishaps, there is likely to be ample warning of any major catastrophe and time for preventive action. Biologists took the initiative in discussing the risks of recombinant DNA and will presumably do so again if the risks prove greater than now anticipated. Only if a risk materializes very suddenly or insidiously will there be no opportunity for countermeasures.

Second, the fate of most organisms released into environments for which they are not adapted is speedy demise. Very occasionally, however, it is the environment that is at a disadvantage, and then the intruding organism runs riot, as has been the case with such foreign imports as the gypsy moth, the fire ant, and the killer bee. The chances of producing a viable new organism accidentally by gene splicing seem rather negligible. Yet the fact is that the technique involves playing evolution's game without fully understanding the rules. Until the rules are better understood, gene splicing inevitably carries the risk of making a forbidden play without knowing what the forfeit will be.

Critics of gene splicing have made much of the irreversibility of accident, of the fact that new organisms cannot be recalled from nature. Perhaps of greater consequence is the irreversibility of the technology. Now that it has been invented, gene splicing and its successor techniques will be a part of industrial civilizations for as long as they endure. More to be feared than accidental mishaps, maybe, are the possibilities for deliberate misuse and abuse.

"Today power can come in very small packages," former CIA Director William Colby told a Senate committee in March 1977: "It no longer has to depend on a great economy and great population. It can come in nuclear form, in biological form, in chemical form, and be available to a more or less reckless despot in some little country who is developing his own homemade backyard atomic bomb."[1]

The point Colby alludes to is that it is sometimes easier to invent a technology than to control it. Even nuclear technology, demanding as it is, now seems so accessible to small countries or even groups that the United States government has determined to write off the millions of dollars invested in the plutonium cycle. The technology of chemical weapons is one that with more forethought might never have been brought into being. The fruits of that folly have not yet been gathered, but, given time, they may be. Through a series of perhaps inevitable indiscretions, the formulas and modes of synthesis of such extraordinarily lethal substances as VX gas are now all but available in the open literature.

Biological warfare is perhaps the most abhorrent of all military arts. That did not prevent the U.S. Army from spending millions of dollars in a twenty-five-year effort to improve upon the unpleasantness of microorganisms noxious to man, and a similar endeavor was undertaken in the Soviet Union. It is fortunate that with the techniques available in the 1950s and 1960s neither side was able to fashion biological weapons of any military significance. In the cosmetic tradition of arms treaties forbidding what neither party intends to do in any case, the two countries ratified their failure in the Biological Weapons Convention, which forbids the development, production, and stockpiling of such weapons.

The United States renounced offensive use of biological weapons in 1969, long before gene splicing was even contemplated. The Convention, which came into force in March 1975, a month after the Asilomar conference, erects significant institutional barriers in the way of military planners in either country who might argue that gene splicing makes it worth resuming a biological warfare program. The military establishment has not forgotten about biological weapons, however. Sometime early in 1977 the Department of Defense asked a government committee on recombinant DNA if in periods of national emergency it could be exempted from the requirement to register all gene-splicing research with the Department of Health, Education, and Welfare. The Defense Department has stated that it is not at present conducting any gene-splicing research (the Convention, it should be noted, does not prohibit research on biological weapons). It is not known whether other government agencies, such as the CIA or the National Security Agency, have recombinant DNA research programs.

Coming at a particularly opportune moment, the Biological Weapons Convention should significantly delay, possibly even prevent, the military exploitation of gene splicing by the United States and Soviet Union and by the sixty or so other nations that have ratified it. Unfortunately, that does not guarantee the technique will never be adapted to military use. As the technology grows in power, so may the temptations it offers to admirals and generals, "reckless despots," and terrorists. If gene therapy ever becomes technically possible, as proponents of the research suggest, so too may the ability to make novel weapons, such as biological agents that distinguish in their specificity between men and women, children and adults, even between people of different races. To a small nation that couldn't manage the technology of nuclear power, or which wished to act secretly, biological weapons might seem to offer certain advantages, especially in any confrontation with a nuclear power. As weapons, biological agents may not yet be particularly effective but they are not in the bow and arrows league either; the United States would certainly consider nuclear retaliation if so attacked.

For the most part, gene splicing seems a generally beneficent and gentle technology; such malign potential as it could have may never be realized. But whatever potential there is for misuse is made more threatening by the ease of access to the technique. Particular applications may be more sophisticated, but the basic technique is extremely easy to perform. The only special material required, restriction enzymes, can be bought for a few dollars or prepared in a couple of days. Most experiments can be conducted in standard biological laboratories equipped with some $50,000 to $200,000 of standard apparatus.

The technique of gene splicing is so much in its infancy that the limits of the art cannot be discerned; yet the variety of life forms thrown up by the random processes of evolution demonstrates the extraordinary plasticity of the material that biologists are now learning how to work. With gene splicing as a philosopher's stone, the experimentalist may one day be able to transmute leaden

species to golden ones or perform whatever other meta-
morphoses take his or her fancy. Even if such dreams of
demiurgy never materialize, an increasingly potent art is
being placed in the hands of every biologist in the world
with access to a standard laboratory. Society can usually
handle a technology so as to reap its benefits and minimize its risks. Less governable is the individual who might
decide to make some unilateral intervention for what he
or she conceives to be the public good.

One of the social problems that lies ahead is that of
genetic engineering. Racist beliefs and the Nazi attempts
to create a master race have given eugenics an odious
reputation. Yet if the historical precedents are laid aside,
the concept may appear to have certain merits. Why leave
man's inheritance to be shaped by the blind forces of
genetic drift or the brute pressures of natural selection,
to the small extent that it may be operative on today's societies? In *Brave New World* Aldous Huxley harped on the
subbreeds of men, designed to remain content in menial
service to the higher castes. But the only humane justification of genetic engineering would be to improve upon
man's inheritance, not to debase it. The concept of progress is a deeply rooted value of industrialized societies:
when a means is found to design indubitable improvements in the intellectual or emotional architecture of the
human mind, would it be true to that value to ignore the
opportunity of improving a scarcely perfect species?

The concept of human genetic engineering is bound
to be resisted on religious and other grounds. Yet, what
is inherently wrong with creating a group of genetically
improved individuals? The idea may not be entirely democratic in inspiration, but human societies are thor-
oughly accustomed to elites and ruling classes whose
claim to superiority is based on—of all things—heredity.

Will it be technically possible to rejig the human gene
set? The gene-splicing technique is only the first step

down a long road. The germ line of higher organisms may prove quite accessible to alteration—for example, by splicing genes onto a virus and using the virus to infect a sperm or egg. On the other hand, animal cells may for evolutionary reasons be specially organized so as to resist this form of viral usurpation. Yet knowledge of how to add genes effectively to an organism may not be permanently elusive. It would be rash to predict that engineering of the human gene set will assuredly be feasible, but it would be equally adventurous to deny the possibility.

[The social acceptability of engineering the human gene set may prove a more formidable obstacle than the technical problems.] George Wald, for example, has already proposed that the human genome should be declared inviolable. But the precept, even if acceptable, is vulnerable to erosion. The advantages of genetic engineering are going to be demonstrated first in the skillful improvement of crop plants and domestic animals. Next will come a development opposed by only Luddites and religious obscurantists, the gene-splice treatment of some of the fifteen hundred human diseases that are now known to be genetically determined. Means of genetic manipulation may then be discovered that enhance the natural process of development and enable each individual to realize his full genetic potential.

Each such advance would surely be as intensely debated as were the first uses of the gene-splicing technique, but the outcome of such debates is seldom in serious doubt: the forces of progress will generally prevail over unsubstantiated forebodings of theoretical hazards. Yet by the time that the human genome has been improved a little, for the best of reasons, there remains no clear barrier against improving it a lot. The dilemma then raised is more than purely taxonomic: a substantial improvement on the human gene set, once we know how to effect it, will produce a creature as different from man as is man from the apes—in other words, a new species.

Such a speculation lies far within the realms of science fiction. Yet someone reviewing the possibility before the invention of gene splicing might well have opined that of all the steps leading up to it, the least plausible is the first. Perhaps the most troubling single aspect of human genetic engineering is that there is maybe nothing inherently wrong with it, provided that a free choice is made at each step along the way. But is a free choice really possible? Will society be stampeded by a technological imperative of what can be done, will be done?

"We must all get used to the idea that biomedical technology makes possible many things we should never do," Leon Kass has remarked in an essay written well before the advent of gene splicing.[2] If the idea is unaccustomed, that is because of the reverence accorded in the liberal political tradition to progress and reason. These values are the roots of the general belief that technological advance should not be resisted and scientific inquiry never impeded. The strength with which they are held derives from the times and occasions when they were threatened by other systems of belief, a confrontation enshrined in the heroic object lessons of Galileo and the victims of Lysenkoism.

To repudiate these values would be a cultural disaster for the societies in which they are instilled. But over the last few decades there has emerged a certain wariness of technology and the perception that particular technological advances can be resisted without opening the gates to mass Luddism. Society can have progress without the SST, agriculture without DDT, even energy without the plutonium cycle. In medicine a growing insistence on the rights of patients has forced a certain conservatism in the introduction of innovations. There may be a technological imperative, but society is not powerless to say no.

Whether society should feel free to extend its power of choice into basic research, the taproot of technological innovation, is another matter. Those who doubt society's ability to handle technical progress wisely have often suggested that the best way to control technology is to control research. The argument is beginning to be raised in the debate about gene splicing, together with the charge that scientists' protests should be discounted because they have a vested interest in pushing ahead with research.

Scientists clearly have a special stake in freedom of inquiry, in which both their livelihood and intellectual interests are deeply involved. But the gene-splicing issue has been openly discussed within the scientific community; not just the strongest defense but also the fiercest criticism of gene splicing continues to come from scientists. Biologists, no less than other citizens, have doubts about genetic engineering and where research may lead, and will presumably continue to voice them. Though researchers do have a conflict of interest where freedom of inquiry is the issue, any divergence between their interests and the public's will quickly become apparent if the precedent of the last few years of open debate is followed. Basic research, moreover, depends on publication, and there is little chance of the public remaining uninformed about any major new advance or its practical applications.

Public authorities have so far shown no interest in the idea that research should be constrained for fear of not being able to deal with its consequences. The controls imposed on gene splicing have been justified solely by concern for safety and not by any ideological consideration.

A different kind of criticism leveled against research relates not to its technological but to its philosophical impact. In the case of molecular biology, the argument is sometimes made that once the purely material nature of

human existence is fully described, it will lower our estimation of ourselves and destroy respect for human values. To quote Kass:

> We have paid some high prices for the technological
> conquest of nature, but none perhaps so high as the intel-
> lectual and spiritual costs of seeing nature as mere material
> for our manipulation, exploitation and transformation.
> With the power for biological engineering now gathering,
> there will be splendid new opportunities for a similar de-
> gradation of our view of man. Indeed, we are already wit-
> nessing the erosion of our idea of man as something splen-
> did or divine, as a creature with freedom and dignity. And
> clearly, if we come to see ourselves as meat, then meat we
> shall become.[3]

Social critics such as Theodore Roszak have made a
major theme of science's corrosive effect on other systems
of values. Science "has taken on the character of a nihil-
istic campaign against the legitimate mysteries of man and
nature," says Roszak.[4] The view of the scientist as a pro-
faner of nature's mysteries is as old as the Romantic re-
bellion:

> *Do not all charms fly*
> *At the mere touch of cold philosophy?*
> *There was an awful rainbow once in heaven:*
> *We know her woof, her texture; she is given*
> *In the dull catalogue of common things.*
> *Philosophy will clip an Angel's wings,*
> *Conquer all mysteries by rule and line,*
> *Empty the haunted air and gnomèd mine—*
> *Unweave a rainbow. . . .*[5]

The price of listening to this particular lament would
have been the industrial revolution; mankind would still
be enjoying the bucolic simplicity that the Romantics idol-
ized. Nevertheless, the discoveries flowing from the gene-
splicing technique will eventually touch on the roots of
human existence and can hardly fail to have an emotional
and intellectual impact of some kind. To read a print-out

of the complete sequence of one's own DNA would probably be a curious experience: no one likes to think of himself or herself as being based on a blueprint that is embodied in a purely chemical system and differs by only a few percent of an admittedly complex formula from some four billion other chemical systems.

It would be a reductionist fallacy to equate a person with his DNA sequence: environmental as well as genetic influences weigh strongly in determining character. All the same, complete understanding of the human gene set, its developmental program, and its differences from the gene sets of other animals, could well affect, and perhaps degrade, humankind's view of itself and its importance in the universal scheme of things. That is no argument for declaring the human gene set off bounds to gene splicers, but it is one of the factors to be considered in the technique's long-term balance sheet, possibly though not necessarily on the debit side.

The ability to manipulate the stuff of life is the ultimate technology. Other technologies are merely extensions of man's hands or mind or senses, serving to amplify or project the capabilities of the user. The further improvement and refinement of these technologies will doubtless continue to be a preoccupation for long into the future. But the impending ability to turn the tools inward for the reshaping of man himself would be an event quite out of the ordinary march of technological progress. Hitherto evolution has seemed as inexorable and irreversible a process as time or entropy; now at least there lies almost within man's grasp a tool for manipulating the force that shaped him, for controlling his own creator.

NOTES

1. Testimony given before the Arms Control Subcommittee of the Senate Committee on Foreign Relations, 16 March, 1977.

2. Leon Kass, "The New Biology: What Price Relieving Man's Estate?" *Science* 174 (19 November, 1971): 779.

3. Leon Kass, "Making Babies—The New Biology and the 'Old' Morality." *The Public Interest,* Winter 1972, p. 53.

4. Theodore Roszak, *Where the Wasteland Ends* (New York: Doubleday, 1972), p. 249.

5. John Keats, *Lamia* II, 229–237.

INDEX